ALICE AND BOB MEET THE WALL OF FIRE

The Biggest Ideas in Science from *Quanta*

edited by Thomas Lin

Quantamagazine

The MIT Press

This book was set in Stone Serif by Westchester Publishing Services. Printed and bound in the United States of America.

Library of Congress Cataloging-in-Publication Data
Names: Lin, Thomas (Journalist), editor.
Title: Alice and Bob meet the wall of fire : the biggest ideas in science
    from Quanta / edited by Thomas Lin ; foreword by Sean Carroll.
Other titles: Quanta science stories
Description: Cambridge, MA : The MIT Press, [2018] | Includes bibliographical
    references and index.
Identifiers: LCCN 2018013315 | ISBN 9780262536349 (pbk. : alk. paper)
Subjects: LCSH: Physics--Popular works. | Biology--Popular works.
Classification: LCC QC24.5 .A445 2018 | DDC 530--dc23 LC record available at
    https://lccn.loc.gov/2018013315

10   9   8   7   6   5   4   3   2   1

Men fear thought as they fear nothing else on earth—more than ruin, more even than death. Thought is subversive and revolutionary, destructive and terrible; thought is merciless to privilege, established institutions, and comfortable habits; thought is anarchic and lawless, indifferent to authority, careless of the well-tried wisdom of the ages. Thought looks into the pit of hell and is not afraid. It sees man, a feeble speck, surrounded by unfathomable depths of silence; yet it bears itself proudly, as unmoved as if it were lord of the universe. Thought is great and swift and free, the light of the world, and the chief glory of man.

—Bertrand Russell in *Why Men Fight*

CONTENTS

V   WHAT MAKES US HUMAN?

VI   HOW DO MACHINES LEARN?

VII   HOW WILL WE LEARN MORE?

VIII WHERE DO WE GO FROM HERE?

FOREWORD

Sean Carroll

Q *uanta* has been a revelation.
All scientists are well aware that diplomatic relations across the border between science and journalism typically range from "wary" to "downright hostile." It's not that there is any natural antipathy between the groups, both of which largely consist of dedicated professionals pursuing an honorable craft. Scientists attempt to discover true things about the natural world and rely on journalists to spread the word of those discoveries to a wide audience; journalists hope to convey those discoveries in engaging and accurate ways and rely on scientists to help them understand and communicate. But the relevant sets of standards and incentives are different. Scientists lapse into jargon in an attempt to preserve precision, and they place enormous value on the precise attribution of credit; journalists want to tell a human story that will be compelling to nonscientists, and they are naturally drawn to the most dramatic angle that can be attached to any particular finding.

Thus, tension. Scientists find themselves insisting that the new way they have developed to slow down the speed of light through a crystal does not, in fact, open any potential routes to warp drive and/or time travel. Journalists have to gently explain that their four-paragraph story won't actually be able to list all 12 coauthors of the original study, nor will there be room to explain the finer points of adiabatic softening. Scientists reluctantly go along with this, only to be aghast when two of the four paragraphs in the final article are devoted to the whimsical antics of the new puppy one of the grad students was bringing to the lab. These are the sacrifices we make for the sake of a human-interest angle.

Quanta was immediately, recognizably, a different kind of science magazine. It explains the real discoveries and speculations at the cutting edge, without implying that they'll make science fiction come true. It lets interesting science be interesting science. I've seen my colleagues' eyes light up when

Quanta is mentioned. "Oh, yeah, that magazine is amazing. They actually talk about science for real!"

It has also made an impact among journalists. I remember conversations with some of my writer friends along the lines of, "I'm really confused about what *Quanta* is looking for. I found a great anecdote about this puppy in a lab, but they seem to care more about what the lab is actually doing. That can't be right, can it?" (The details are entirely fabricated, but the spirit of the conversation is accurate.)

And the *Quanta* approach will, I predict, make an impact on you as you read the pieces in this collection. It's not that the stories are dry or devoid of human interest—quite the contrary. Scientists are people, and their hopes and fears come through vividly in these accounts. But the science is always paramount. And *Quanta* doesn't apologize for that.

Such science it is! I'll confess my bias here, as this collection shines a spotlight on the kind of challenging, speculative, cutting-edge physics that is my own primary interest. When reading through these stories, I get the warm feeling that these are my people.

It's an interesting time for fundamental physics. We have theories of surpassing rigor and beauty that also pass every experimental test—quantum mechanics, Einstein's general relativity for gravity, the Standard Model of particle physics and the Hot Big Bang cosmological model. You'd think we would be happy. But at the same time, we know that these theories are not the final story. They don't play well together, and they leave a number of crucial questions unanswered. How should we move forward in the absence of direct, actionable experimental clues?

In cosmology, researchers have become increasingly intrigued by the notion that we live in a multiverse. That's not quite as philosophical as it sounds. What cosmologists actually mean by a "multiverse" is really a single universe, but one composed of multiple mutually inaccessible regions, in which the conditions are very different from those of all the others. It's not a dorm-room fantasy; this conception of a multiverse arises naturally from inflationary cosmology and string theory, two of the most popular (though still speculative) ideas on the fundamental physics market today.

In these pages you'll read why physicists have been led to contemplate such extreme possibilities. It's not simply that cosmologists got bored and started positing a billion other universes; there are features of our universe, here and now, observed in the lab, that might best be explained if we imagine that we live in just one of a large ensemble of universes. What's more, you'll come to appreciate that our apparent inability to observe a multiverse might not be as cut-and-dried as it seems. This is how science progresses: Researchers take on a tough problem, propose some outlandish-sounding solutions and then work to wrestle those grandiose ideas into a down-and-dirty confrontation with the data.

Not everyone is happy with the multiverse idea, of course. Plenty of working scientists are concerned that the concept is a step too far away from the real work of science, venturing into a realm of unfettered speculation, and the tensions are well documented here. Take a peek inside a gathering of some of the world's brightest minds, where banners are raised in a "battle for the heart and soul of physics." (Human interest!)

Besides the structure of the universe as a whole, a major preoccupation of modern physics is the nature of space-time. You might think this was figured out a century back, when Einstein put forward his general theory of relativity, according to which space and time are unified into a dynamical, four-dimensional space-time whose twists and turns are observable to us as the force of gravity. That seems to be true at the "classical" level—the setup for physics worked out by Isaac Newton centuries ago, in which objects have definite positions and velocities. But now we know about quantum mechanics, which describes the world very differently. According to quantum mechanics, the observable world doesn't march forward with clockwork predictability; the best we can do in any particular situation is to calculate the probability of observing various possible outcomes of an experiment.

The problem—and it's a big one—is that this quantum approach has not yet been reconciled with the curved space-time of general relativity. That's the issue of quantum gravity, which forms a major theme of the pieces collected here. In the title piece, "Alice and Bob Meet the Wall of Fire," we are introduced to the firewall paradox: According to controversial recent research, quantum mechanics implies that observers falling into a black hole will be incinerated by a wall of fire, rather than ultimately being pulled apart by the gravitational field. This distasteful conclusion has led to some outlandish proposed solutions, up to and including the notion that particles that share quantum entanglement may also be associated with wormholes in space-time. Entanglement, indeed, may ultimately be responsible for the existence of space-time itself.

All this talk of entanglement reminds us of a lurking worry when we gird our loins to tackle the difficult problem of quantum gravity: If we're honest about it, we don't really understand quantum mechanics itself, even without dragging gravity into the picture. There's no question that we can *use* quantum mechanics, both to make staggeringly precise predictions and to uncover the hidden beauty in the dynamics of particles and fields. But ask physicists what quantum mechanics actually *is* and they'll look nervous and discomfited. People have ideas, of course, some of which you will meet here: Maybe they will find hidden variables like the kind Einstein and others dreamed of long ago, or perhaps they can build the entire edifice from scratch.

Or perhaps the secret to reconciling quantum theory with space-time isn't a better understanding of the quantum, but a better understanding of time. Maybe quantum effects are crucial to understanding the prosaic-seeming question of why time moves forward and not backward. Or, even more dramatically, perhaps entanglement is at the heart of what time itself really is.

All of this might seem like an overabundance of *perhapses* and *maybes*. Isn't science supposed to be a repository of firm, established truths about nature?

Eventually, yes. But the process of getting there is messy and unpredictable, full of false starts and discarded hypotheses. The focus in *Quanta* is on science at the cutting edge, where things are never simple. You can see the dialectical process in action in "A Debate over the Physics of Time." It's an up-close view of physicists and philosophers coming together to exchange ideas and occasionally shout at each other ("I'm sick and tired of this block universe!"—yes, we really talk like that).

The science in this book is by no means confined to physics. Among all the sciences, physics seems hard because it is actually quite simple. That is, we can take a basic system like a rocking pendulum or a planet orbiting the sun and explain more or less exactly how it behaves. Having accomplished this, we move on to harder things—curved space-time, subatomic particles, the origin of the universe itself. A lot of the low-hanging fruit has already been picked, leaving cutting-edge research to concern itself with some extremely ambitious ideas.

The rest of science is messier and, correspondingly, harder. In biology, we can take a basic system like an earthworm or even a bacterium, and—well, suffice it to say, we are nowhere close to offering up an exact description of how these things work. Physics leans heavily on the fact that we can often ignore large sets of complications (friction, noise, air resistance) and still get a pretty good answer. In the wild and woolly world of biological complexity, all of the pieces matter.

In *Quanta*'s takes on biology and life, a couple of themes emerge. One, unsurprisingly, is Darwin's theory of evolution by natural selection. Perhaps more surprising is how much richer the paradigm has become as scientists continue to tinker with it. Darwin himself, of course, knew nothing of microbiology or DNA, and modern scientists are racing to improve our understanding of how genetic information is passed down through generations. And in a game-changing discovery whose ultimate implications are still far from understood, a tool called CRISPR promises to allow us to dig into the basic letters of the genetic code to edit DNA directly. The CRISPR piece surely won't be the last story about gene editing to appear in *Quanta*— it holds the promise to revolutionize medicine and perhaps exert profound changes on what it means to be human.

Another theme is the origin of life and its complexity. One might worry that life seems paradoxical when considered in the broader sweep of cosmic history. According to the second law of thermodynamics, entropy (a measure of disorder or randomness) increases over time; how, then, did something as orderly and nonrandom as living beings come into existence? An intriguing modern view is that this isn't a paradox at all, but rather a natural connection: Perhaps life began not despite the increase in entropy but because of it. After all, although an individual organism may be intricately ordered, it will inevitably increase the entropy of the universe by the simple fact that it metabolizes food to stay alive. But we would still like a detailed, historical understanding of why life became complex in the particular ways that it did—it seems possible that this is a process that happens even if natural selection doesn't nudge us in that direction. Even structures as specific as neurons may have independently come into existence more than once.

Among all the leaves on the tree of life, the one that many of us are most curious about is humanity itself. Evolution, needless to say, plays a central role in explaining how we became who we are. *Homo sapiens* didn't burst into existence fully formed, like Athena from the brow of Zeus; our plucky ability to thrive in a diverse set of environments may be a legacy of the genes we inherited from our prehistoric ancestors. What truly set us apart, of course, was the "brain boom" that began 3 million years ago, when the human brain began to almost quadruple in size in comparison to that of its predecessors.

The quest to understand how that brain works will doubtless keep scientists occupied for many generations to come. One step is simply to understand how our mind's various capacities are assembled over the course of our early lives. It happens rapidly: Within six months after birth, an infant's brain has the same basic organization as that of an adult. Once it's assembled, as marvelous as it is, the brain falls far short of being a perfectly rational machine. If you feel a twinge of guilt at your inability to resist another slice of pizza or an extra scoop of ice cream, take solace in the fact that these tendencies are simply part of a neurological optimization strategy, as the brain saves its energy for more important tasks. Even something as fundamentally human as the tendency to get lonely can be given a scientific explanation, as a type of incentive to inspire us to social cooperation.

Brains matter because we use them to think. What is this "thinking" on which we place such value? For many years, scientists and philosophers have speculated that something akin to human thought could also conceivably take place inside a mechanical device such as a computer. We may be entering the first generation in which this prediction will be put to the test. Along some fronts, progress has been impressive: Model "brains" built from tiny atomic switches can learn new things, and artificial agents can display something recognizable as curiosity. Computers nowadays handily defeat the world's best chess players, and recent victories at the even more complex game of Go hint at the development of something that might even be labeled "intuition." A great deal of excitement has accompanied the rise of "deep learning," a method of allowing networks of artificial neurons to train themselves to recognize and manipulate high-level concepts. Despite some initial successes, there is a lingering worry that after such a network trains itself, its creators can't understand how it is actually doing what it does. So begins the quest to construct a theory of deep learning.

Surveying the scope and depth of this collection of articles, one cannot help but feel optimistic about the future—the future of not only science but also science journalism.

In neither case is the optimism pure and unadulterated. On the science side, we must always be prepared for those times when our experiments don't give us the answers we were hoping for. That's happened in recent years in particle physics, where the lack of new particles to be found beyond the Higgs boson has left physicists scratching their heads and wondering whether there's a fundamental mistake in how we've been thinking about nature. But for every disappointment, there are multiple triumphs: The detection of gravitational waves from merging black holes and neutron stars has sparked a revolution that promises to energize astrophysics for years to come. Whatever our frustrations may be with the current pace of scientific progress, we must remember that it's not the universe's job to keep us happy; it's our job to do the best we can at uncovering its secrets.

At the same time, science journalism has suffered from the general downturn in traditional media coverage, including an elimination of many staff jobs at newspapers and magazines. Reason for hope is to be found in a broadening ecosystem as a wide variety of outlets have sprung into existence. Within this group, *Quanta* is a guiding light, providing other media with an example of science writing at its best.

In a sense, these stories are a snapshot of a moment in time; the thing about the cutting edge is that it keeps moving. Some of the scientific ideas explored here will fade away or be dramatically ruled out whereas others will become absolutely central to how we think about the universe. However, the ideal of communicating science by taking it seriously, by wrestling with difficult concepts and explaining them in honest, clear language—that is here to stay.

INTRODUCTION

Thomas Lin
Quanta Magazine editor in chief

I t's hard to beat a good science or math story.

Take the events of July 4, 2012. That morning, scientists at the world's biggest physics experiment, the Large Hadron Collider near Geneva, Switzerland, made the biggest announcement of their lives. After investing 20 years in the effort to design and construct the LHC, only to suffer a demoralizing malfunction soon after it went online in 2008, they had finally discovered the Higgs boson, a particle without which life and the universe as we know it could not exist. The next year, Peter W. Higgs and François Englert, whose theoretical work in the 1960s predicted the Higgs particle, won the Nobel Prize in Physics.

A year later, the documentary film *Particle Fever* chronicled the hopes and dreams of the thousands of researchers behind the Higgs discovery. In one of my favorite scenes, the theorist David Kaplan is shown in 2008 explaining to a packed lecture hall why they built what the experimentalist Monica Dunford calls "a five-story Swiss watch." Unmoved by Kaplan's talk, an economist demands to know what people stand to gain from the multibillion-dollar experiment: "What's the economic return? How do you justify all this?"

"I have no idea," Kaplan answers bluntly. He must get this question all the time. Patiently, he explains that big breakthroughs in basic science "occur at a level where you're not asking, 'What is the economic gain?' You're asking, 'What do we not know, and where can we make progress?'" In their purest forms, science and mathematics are not about engineering practical applications or cashing in on them, though those things often do happen later, sometimes much later. They're about learning something you didn't know before.

"So, what is the LHC good for?" Kaplan asks the economist, setting up the death blow: "It could be nothing, other than just—understanding everything."

As it happens, this book picks up where *Particle Fever* leaves off in telling the story of the quest to understand everything. The renowned theoretical physicist Nima Arkani-Hamed has described such efforts in fundamental physics as "trying to understand, in the simplest possible way, the smallest set of basic principles from which everything else in principle follows." The fewer the assumptions, the approximations, the contortions—or so the thinking goes—the closer we are to the truth. The Higgs boson has been discovered and the Standard Model of particle physics is now complete. The problem is, absent new particles beyond the Standard Model, the universe doesn't make sense. How, then, are we to make sense of it?

In *Alice and Bob Meet the Wall of Fire: The Biggest Ideas in Science from* Quanta, and *The Prime Number Conspiracy: The Biggest Ideas in Math from* Quanta, we join some of the greatest scientific and mathematical minds as they test the limits of human knowledge. The stories presented in these two companion volumes reveal efforts over the past five years or so to untangle the mysteries of the universe—its origins and basic laws, its contents big and small, its profoundly complex living inhabitants—and to unlock the universal language of nature. They penetrate the big questions and uncover some of the best ideas and theories for understanding our physical, biotic and logical worlds. Meanwhile, they illuminate the essential issues under debate as well as the obstacles hindering further progress.

In selecting and editing *Quanta Magazine* articles for these volumes, I tried to venture beyond the usual mixtape format of "best of" anthologies and greatest hits compilations. Instead, I wanted to send readers on breathtaking intellectual journeys to the bleeding edge of discovery strapped to the narrative rocket of humanity's never-ending pursuit of knowledge. But what might those excursions actually look like? These nonfiction adventures, it turns out, explore core questions about the essence of prime numbers, whether our universe is "natural," the natures of time and infinity, our strange quantum reality, whether space-time is fundamental or emergent, the insides and outsides of black holes, the origin and evolution of life, what makes us human, the hopes for and limitations of computing, the role of mathematics in science and society, and just where these questions are taking us. The stories in these books reveal how cutting-edge research is done—how the productive tension between theory, experiment and mathematical intuition, through triumphs, failures and null results, cuts a path forward.

What is *Quanta*? Albert Einstein called photons "quanta of light." *Quanta Magazine* strives to illuminate the dark recesses in science and mathematics where some of the most provocative and fundamental ideas are cultivated out of public view. Not that anyone is trying to hide them. The work hides in plain sight at highly technical conferences and workshops, on the pre-print site arxiv.org and in impenetrable academic journals. These are not easy subjects to understand, even for experts in adjacent fields, so it's not surprising that only Higgs-level discoveries are widely covered by the popular press.

The story of *Quanta* began in 2012, just weeks after the Higgs announcement. With the news industry still reeling from the 2008 financial crisis and secular declines in print advertising, I had the not-so-brilliant idea to start a science magazine. The magazine I envisioned would apply the best editorial standards of publications like the *New York Times* and the *New Yorker*, but its coverage would differ radically from that of existing news outlets. For one thing, it wouldn't report on anything you might actually find useful. This magazine would not publish health or medical news or breathless coverage of the latest technological breakthroughs. There would be no advice on which foods or vitamins to consume or avoid, which exercises to wedge into your day, which gadgets are must-buys. No stories about crumbling infrastructure or awesome feats of engineering. It wouldn't even keep you updated about the latest NASA mission, exoplanet find or SpaceX rocket launch. There's nothing wrong with any of this, of course. When accurately reported, deftly written, and carefully fact checked, it's "news you can use." But I had other ideas. I wanted a science magazine that helps us achieve escape velocity beyond our own small worlds but is otherwise useless in the way the LHC is useless. This useless magazine became *Quanta*.

My colleagues and I also treat our readers differently. We don't protect them from the central concepts or from the process of how new ideas come to be. Indeed, the ridiculously difficult science and math problems and the manner in which an individual or collaboration makes progress serve as the very conflicts and resolutions that drive *Quanta* narratives. We avoid jargon, but we don't protect readers from the science itself. We trust readers, whether they have a science background or not, to be intellectually curious enough to want to know more, so we give you more.

Like the magazine, this book is for anyone who wants to understand how nature works, what the universe is made of, and how life got its start and evolved into its myriad forms. It's for curiosity seekers who want a front-row seat for breakthroughs to the biggest mathematical puzzles and whose idea of fun is to witness the expansion of our mathematical universe.

If I may offer an adaptation of Shel Silverstein's famous lyric invitation (my sincere apologies to the late Mr. Silverstein):

> If you are a dreamer, come in,
> If you are a dreamer, a thinker, a curiosity seeker,
> A theorizer, an experimenter, a *mathematiker* …
> If you're a tinkerer, come fill my beaker
> For we have some mind-bendin' puzzles to examine.
> Come in!
> Come in!

WHY DOESN'T OUR UNIVERSE MAKE SENSE?

IS NATURE UNNATURAL?

Natalie Wolchover

O n an overcast afternoon in late April 2013, physics professors and students crowded into a wood-paneled lecture hall at Columbia University for a talk by Nima Arkani-Hamed, a high-profile theorist visiting from the Institute for Advanced Study in nearby Princeton, New Jersey. With his dark, shoulder-length hair shoved behind his ears, Arkani-Hamed laid out the dual, seemingly contradictory implications of recent experimental results at the Large Hadron Collider in Europe.

"The universe is inevitable," he declared. "The universe is impossible."

The spectacular discovery of the Higgs boson in July 2012 confirmed a nearly 50-year-old theory of how elementary particles acquire mass, which enables them to form big structures such as galaxies and humans. "The fact that it was seen more or less where we expected to find it is a triumph for experiment, it's a triumph for theory, and it's an indication that physics works," Arkani-Hamed told the crowd.

However, in order for the Higgs boson to make sense with the mass (or equivalent energy) it was determined to have, the LHC needed to find a swarm of other particles, too. None turned up.

With the discovery of only one particle, the LHC experiments deepened a profound problem in physics that had been brewing for decades. Modern equations seem to capture reality with breathtaking accuracy, correctly predicting the values of many constants of nature and the existence of particles like the Higgs. Yet a few constants—including the mass of the Higgs boson—are exponentially different from what these trusted laws indicate they should be, in ways that would rule out any chance of life, unless the universe is shaped by inexplicable fine-tunings and cancellations.

In peril is the notion of "naturalness," Albert Einstein's dream that the laws of nature are sublimely beautiful, inevitable and self-contained. Without it, physicists face the harsh prospect that those laws are just an arbitrary, messy outcome of random fluctuations in the fabric of space and time.

In papers, talks and interviews, Arkani-Hamed and many other top physicists are confronting the possibility that the universe might be unnatural. (There is wide disagreement, however, about what it would take to prove it.)

"Ten or 20 years ago, I was a firm believer in naturalness," said Nathan Seiberg, a theoretical physicist at the Institute, where Einstein taught from 1933 until his death in 1955. "Now I'm not so sure. My hope is there's still something we haven't thought about, some other mechanism that would explain all these things. But I don't see what it could be."

Physicists reason that if the universe is unnatural, with extremely unlikely fundamental constants that make life possible, then an enormous number of universes must exist for our improbable case to have been realized. Otherwise, why should we be so lucky? Unnaturalness would give a huge lift to the multiverse hypothesis, which holds that our universe is one bubble in an infinite and inaccessible foam. According to a popular but polarizing framework called string theory, the number of possible types of universes that can bubble up in a multiverse is around 10^{500}. In a few of them, chance cancellations would produce the strange constants we observe.

In such a picture, not everything about this universe is inevitable, rendering it unpredictable. Edward Witten, a string theorist at the Institute, said by email, "I would be happy personally if the multiverse interpretation is not correct, in part because it potentially limits our ability to understand the laws of physics. But none of us were consulted when the universe was created."

"Some people hate it," said Raphael Bousso, a physicist at the University of California at Berkeley who helped develop the multiverse scenario. "But I just don't think we can analyze it on an emotional basis. It's a logical possibility that is increasingly favored in the absence of naturalness at the LHC."

What the LHC does or doesn't discover in future runs is likely to lend support to one of two possibilities: Either we live in an overcomplicated but stand-alone universe, or we inhabit an atypical bubble in a multiverse. "We will be a lot smarter five or 10 years from today because of the LHC," Seiberg said. "So that's exciting. This is within reach."

COSMIC COINCIDENCE

Einstein once wrote that for a scientist, "religious feeling takes the form of a rapturous amazement at the harmony of natural law" and that "this feeling is the guiding principle of his life and work." Indeed, throughout the 20th century, the deep-seated belief that the laws of nature are harmonious—a belief in "naturalness"—has proven a reliable guide for discovering truth.

"Naturalness has a track record," Arkani-Hamed told *Quanta*. In practice, it is the requirement that the physical constants (particle masses and other fixed properties of the universe) emerge directly from the laws of physics, rather than resulting from improbable cancellations. Time and again, whenever a constant appeared fine-tuned, as if its initial value had been magically dialed to offset other effects, physicists suspected they were missing something. They would seek and inevitably find some particle or feature that materially dialed the constant, obviating a fine-tuned cancellation.

This time, the self-healing powers of the universe seem to be failing. The Higgs boson has a mass of 126 giga-electron-volts, but interactions with the other known particles should add about 10,000,000,000,000,000,000 giga-electron-volts to its mass. This implies that the Higgs' "bare mass," or starting value before other particles affect it, just so happens to be the negative of that astronomical number, resulting in a near-perfect cancellation that leaves just a hint of Higgs behind: 126 giga-electron-volts.

Physicists have gone through three generations of particle accelerators searching for new particles, posited by a theory called supersymmetry, that would drive the Higgs mass down exactly as much as the known particles drive it up. But so far they've come up empty-handed.

At this point, even if new particles are found at the LHC, they will almost definitely be too heavy to influence the Higgs mass in quite the right way. Physicists disagree about whether this is acceptable in a natural, stand-alone universe. "Fine-tuned a little—maybe it just happens," said Lisa Randall, a professor at Harvard University. But in Arkani-Hamed's opinion, being "a little bit tuned is like being a little bit pregnant. It just doesn't exist."

If no new particles appear and the Higgs remains astronomically fine-tuned, then the multiverse hypothesis will stride into the limelight. "It doesn't mean it's right," said Bousso, a longtime supporter of the multiverse picture, "but it does mean it's the only game in town."

A few physicists—notably Joe Lykken of Fermi National Accelerator Laboratory in Batavia, Illinois, and Alessandro Strumia of the University of Pisa in Italy—see a third option. They say that physicists might be misgauging the effects of other particles on the Higgs mass and that when calculated differently, its mass appears natural. This "modified naturalness" falters when additional particles, such as the unknown constituents of dark matter, are included in calculations—but the same unorthodox path could yield other ideas.[1] "I don't want to advocate, but just to discuss the consequences,"

Strumia said during a 2013 talk at Brookhaven National Laboratory in Long Island, New York.

However, modified naturalness cannot fix an even bigger naturalness problem that exists in physics: the fact that the cosmos wasn't instantly annihilated by its own energy the moment after the Big Bang.

DARK DILEMMA

The energy built into the vacuum of space (known as vacuum energy, dark energy or the cosmological constant) is a baffling trillion trillion trillion trillion trillion trillion trillion trillion trillion times smaller than what is calculated to be its natural, albeit self-destructive, value. No theory exists about what could naturally fix this gargantuan disparity. But it's clear that the cosmological constant has to be enormously fine-tuned to prevent the universe from rapidly exploding or collapsing to a point. It has to be fine-tuned in order for life to have a chance.

To explain this absurd bit of luck, the multiverse idea has been growing mainstream in cosmology circles over the past few decades. It got a credibility boost in 1987 when the Nobel Prize-winning physicist Steven Weinberg, now a professor at the University of Texas at Austin, calculated that the cosmological constant of our universe is expected in the multiverse scenario.[2] Of the possible universes capable of supporting life—the only ones that can be observed and contemplated in the first place—ours is among the least fine-tuned. "If the cosmological constant were much larger than the observed value, say by a factor of 10, then we would have no galaxies," explained Alexander Vilenkin, a cosmologist and multiverse theorist at Tufts University. "It's hard to imagine how life might exist in such a universe."

Most particle physicists hoped that a more testable explanation for the cosmological constant problem would be found. None has. Now, physicists say, the unnaturalness of the Higgs makes the unnaturalness of the cosmological constant more significant. Arkani-Hamed thinks the issues may even be related. "We don't have an understanding of a basic extraordinary fact about our universe," he said. "It is big and has big things in it."

The multiverse turned into slightly more than just a hand-waving argument in 2000, when Bousso and Joseph Polchinski, a professor of theoretical physics at the University of California at Santa Barbara, found a mechanism that could give rise to a panorama of parallel universes. String theory, a hypothetical "theory of everything" that regards particles as invisibly small vibrating lines, posits that space-time is 10-dimensional. At the human scale, we experience just three dimensions of space and one of time, but string theorists argue that six extra dimensions are tightly knotted at every point in the fabric of our 4-D reality. Bousso and Polchinski calculated that there are around 10^{500} different ways for those six dimensions to be knotted (all tying up varying amounts of energy), making an inconceivably vast and diverse array of universes possible.[3] In other words, naturalness is not required. There isn't a single, inevitable, perfect universe.

"It was definitely an aha-moment for me," Bousso said. But the paper sparked outrage.

"Particle physicists, especially string theorists, had this dream of predicting uniquely all the constants of nature," Bousso explained. "Everything would just come out of math and pi and twos [or other simple constants]. And we came in and said, 'Look, it's not going to happen, and there's a reason it's not going to happen. We're thinking about this in totally the wrong way.'"

LIFE IN A MULTIVERSE

The Big Bang, in the Bousso-Polchinski multiverse scenario, is a fluctuation. A compact, six-dimensional knot that makes up one stitch in the fabric of reality suddenly shape-shifts, releasing energy that forms a bubble of space and time. The properties of this new universe are determined by chance: the amount of energy unleashed during the fluctuation. The vast majority of universes that burst into being in this way are thick with vacuum energy; they either expand or collapse so quickly that life cannot arise in them. But some atypical universes, in which an improbable cancellation yields a tiny value for the cosmological constant, are much like ours.

In a paper posted in 2013 to the physics preprint website arXiv.org, Bousso and a Berkeley colleague, Lawrence Hall, argued that the Higgs mass makes sense in the multiverse scenario, too.[4] They found that bubble universes that contain enough visible matter (compared to dark matter) to support life most often have supersymmetric particles beyond the energy range of the LHC, and a fine-tuned Higgs boson. Similarly, other physicists showed in 1997 that if the Higgs boson were five times heavier than it is, this would suppress the formation of atoms other than hydrogen, resulting, by yet another means, in a lifeless universe.[5]

Despite these seemingly successful explanations, many physicists worry that there is little to be gained by adopting the multiverse worldview. Parallel universes cannot be tested for; worse, an unnatural universe resists understanding. "Without naturalness, we will lose the motivation to look for new physics," said Kfir Blum in 2013, when he was a physicist at the

Institute for Advanced Study. "We know it's there, but there is no robust argument for why we should find it." That sentiment is echoed again and again: "I would prefer the universe to be natural," Randall said.

But theories can grow on physicists. After spending more than a decade acclimating himself to the multiverse, Arkani-Hamed now finds it plausible—and a viable route to understanding the ways of our world. "The wonderful point, as far as I'm concerned, is basically any result at the LHC will steer us with different degrees of force down one of these divergent paths," he said. "This kind of choice is a very, very big deal."

Naturalness could pull through. Or it could be a false hope in a strange but comfortable pocket of the multiverse.

As Arkani-Hamed told the audience at Columbia University, "stay tuned."

ALICE AND BOB MEET THE WALL OF FIRE

Jennifer Ouellette

A lice and Bob, beloved characters of various thought experiments in quantum mechanics, are at a crossroads. The adventurous, rather reckless Alice jumps into a very large black hole, leaving a presumably forlorn Bob outside the event horizon—a black hole's point of no return, beyond which nothing, not even light, can escape.

Conventionally, physicists have assumed that if the black hole is large enough, Alice won't notice anything unusual as she crosses the horizon. In this scenario, colorfully dubbed "No Drama," the gravitational forces won't become extreme until she approaches a point inside the black hole called the singularity. There, the gravitational pull will be so much stronger on her feet than on her head that Alice will be "spaghettified."

Now a new hypothesis is giving poor Alice even more drama than she bargained for. If this alternative is correct, as the unsuspecting Alice crosses the event horizon, she will encounter a massive wall of fire that will incinerate her on the spot. As unfair as this seems for Alice, the scenario would also mean that at least one of three cherished notions in theoretical physics must be wrong.

When Alice's fiery fate was proposed in the summer of 2012, it set off heated debates among physicists, many of whom were highly skeptical. "My initial reaction was, 'You've got to be kidding,'" admitted Raphael Bousso. He thought a forceful counterargument would quickly emerge and put the matter to rest. Instead, after a flurry of papers debating the subject, he and his colleagues realized that this had the makings of a mighty fine paradox.

THE "MENU FROM HELL"

Paradoxes in physics have a way of clarifying key issues. At the heart of this particular puzzle lies a conflict between three fundamental postulates beloved by many physicists. The first, based on the equivalence principle of general relativity, leads to the No Drama scenario: Because Alice is in free fall as she crosses the horizon, and there is no difference between free fall and inertial motion, she shouldn't feel extreme effects of gravity. The second postulate is unitarity, the assumption, in keeping with a fundamental tenet of quantum mechanics, that information that falls into a black hole is not irretrievably lost. Lastly, there is what might be best described as "normality," namely, that physics works as expected far away from a black hole even if it breaks down at some point within the black hole—either at the singularity or at the event horizon.

Together, these concepts make up what Bousso ruefully calls "the menu from hell." To resolve the paradox, one of the three must be sacrificed, and nobody can agree on which one should get the ax.

Physicists don't lightly abandon time-honored postulates. That's why so many find the notion of a wall of fire downright noxious. "It is odious," John Preskill of the California Institute of Technology declared in December 2012 at an informal workshop organized by Stanford University's Leonard Susskind. For two days, 50 or so physicists engaged in a spirited brainstorming session, tossing out all manner of crazy ideas to try to resolve the paradox, punctuated by the rapid-fire *tap-tap-tap* of equations being scrawled on a blackboard. But despite the collective angst, even the firewall's fiercest detractors have yet to find a satisfactory solution to the conundrum.

According to the string theorist Joseph Polchinski, who spoke to *Quanta* soon after making the proposal and died in February 2018 of brain cancer, the simplest solution is that the equivalence principle breaks down at the event horizon, thereby giving rise to a firewall. Polchinski was a co-author of the paper that started it all, along with Ahmed Almheiri, Donald Marolf and James Sully—a group often referred to as "AMPS."[1] Even Polchinski thought the idea was a little crazy. It's a testament to the knottiness of the problem that a firewall is the least radical potential solution.

If there is an error in the firewall argument, the mistake is not obvious. That's the hallmark of a good scientific paradox. And it comes at a time when theorists are hungry for a new challenge: The Large Hadron Collider has failed to turn up any data hinting at exotic physics beyond the Standard Model. "In the absence of data, theorists thrive on paradox," Polchinski quipped.

If AMPS is wrong, according to Susskind, it is wrong in a really interesting way that will push physics forward, hopefully toward a robust theory of quantum gravity.[2] Black holes are interesting to physicists, after all, because both general relativity and quantum mechanics can apply, unlike in the rest of the universe, where objects are governed by quantum mechanics at the subatomic scale and by general relativity on the macroscale. The two "rule books" work well enough in their respective regimes, but physicists would love to combine them to shed light on anomalies like black holes and, by extension, the origins of the universe.

AN ENTANGLED PARADOX

The issues are complicated and subtle—if they were simple, there would be no paradox—but a large part of the AMPS argument hinges on the notion of monogamous quantum entanglement: You can only have one kind of entanglement at a time. AMPS argues that two different kinds of entanglement are needed in order for all three postulates on the "menu from hell" to be true. Since the rules of quantum mechanics don't allow you to have both entanglements, one of the three postulates must be sacrificed.

Entanglement—which Albert Einstein ridiculed as "spooky action at a distance"—is a well-known feature of quantum mechanics (in the thought experiment, Alice and Bob represent an entangled particle pair). When subatomic particles collide, they can become invisibly connected, though they may be physically separated. Even at a distance, they are inextricably interlinked and act like a single object. So knowledge about one partner can instantly reveal knowledge about the other. The catch is that you can only have one entanglement at a time.

Under classical physics, as Preskill explained on Caltech's Quantum Frontiers blog, Alice and Bob can both have copies of the same newspaper, which gives them access to the same information. Sharing this bond of sorts makes them "strongly correlated." A third person, "Carrie," can also buy a copy of that newspaper, which gives her equal access to the information it contains, thereby forging a correlation with Bob without weakening his correlation with Alice. In fact, any number of people can buy a copy of that same newspaper and become strongly correlated with one another.

But with quantum correlations, that is not the case. For Bob and Alice to be maximally entangled, their respective newspapers must have the same orientation, whether right side up, upside down or sideways. So long as the orientation is the same, Alice and Bob will have access to the same information. "Because there is just one way to read a classical newspaper and lots of ways to read a quantum newspaper, the quantum correlations are stronger than the classical ones," Preskill said. That makes it impossible for Bob to become as strongly entangled with Carrie as he is with Alice without sacrificing some of his entanglement with Alice.

This is problematic because there is more than one kind of entanglement associated with a black hole, and under the AMPS hypothesis, the two come into conflict. There is an entanglement between Alice, the in-falling observer, and Bob, the outside observer, which is needed to preserve No Drama. But there is also a second entanglement that emerged from another famous paradox in physics, one related to the question of whether information is lost in a black hole. In the 1970s, Stephen Hawking realized that black holes aren't completely black. While nothing might seem amiss to Alice as she crosses the event horizon, from Bob's perspective, the horizon would appear to be glowing like a lump of coal—a phenomenon now known as Hawking radiation.

This radiation results from virtual particle pairs popping out of the quantum vacuum near a black hole. Normally they would collide and annihilate into energy, but sometimes one of the pair is sucked into the black hole while the other escapes to the outside world. The mass of the black hole, which must decrease slightly to counter this effect and ensure that energy is still conserved, gradually winks out of existence. How fast it evaporates depends on the black hole's size: The bigger it is, the more slowly it evaporates.

Hawking assumed that once the radiation evaporated altogether, any information about the black hole's contents contained in that radiation would be lost. "Not only does God play dice, but he sometimes confuses us by throwing them where they can't be seen," he famously declared. He and the Caltech physicist Kip Thorne even made a bet with a dubious Preskill in the 1990s about about whether or not information is lost in a black hole. Preskill insisted that information must be conserved; Hawking and Thorne believed that information would be lost. Physicists eventually realized that it is possible to preserve the information at a cost: As the black hole evaporates, the Hawking radiation must become increasingly entangled with the area outside the event horizon. So when Bob observes that radiation, he can extract the information.

But what happens if Bob were to compare his information with Alice's after she has passed beyond the event horizon? "That would be disastrous," Bousso explained, "because Bob, the outside observer, is seeing the same information in the Hawking radiation, and if they could talk about it, that would be quantum Xeroxing, which is strictly forbidden in quantum mechanics."

Physicists, led by Susskind, declared that the discrepancy between these two viewpoints of the black hole is fine so long as it is impossible for Alice and Bob to share their respective information. This concept, called complementarity, simply holds that there is no direct contradiction because no single observer can ever be both inside and outside the event horizon. If Alice crosses the event horizon, sees a star inside that radius and wants to tell Bob about it, general relativity has ways of preventing her from doing so.

Susskind's argument that information could be recovered without resorting to quantum Xeroxing proved convincing enough that Hawking conceded his bet with Preskill in 2004, presenting the latter with a baseball encyclopedia from which, he said, "information can be retrieved at will." But perhaps Thorne, who refused to concede, was right to be stubborn.

Bousso thought complementarity would come to the rescue yet again to resolve the firewall paradox. He soon realized that it was insufficient. Complementarity is a theoretical concept developed to address a specific problem, namely, reconciling the two viewpoints of observers inside and outside the event horizon. But the firewall is just the tiniest bit outside the event horizon, giving Alice and Bob the same viewpoint, so complementarity won't resolve the paradox.

TOWARD QUANTUM GRAVITY

If they wish to get rid of the firewall and preserve No Drama, physicists need to find a new theoretical insight tailored to this unique situation or concede that perhaps Hawking was right all along, and information is indeed lost, meaning Preskill might have to return his encyclopedia. So it was surprising to find Preskill suggesting that his colleagues at the Stanford workshop at least reconsider the possibility of information loss. Although we don't know how to make sense of quantum mechanics without unitarity, "that doesn't mean it can't be done," he said. "Look in the mirror and ask yourself: Would I bet my life on unitarity?"

Polchinski argued persuasively in 2012 that you need Alice and Bob to be entangled to preserve No Drama, and you need the Hawking radiation to be entangled with the area outside the event horizon to conserve quantum information. But you can't have both. If you sacrifice the entanglement of the Hawking radiation with the area outside the event horizon, you lose information. If you sacrifice the entanglement of Alice and Bob, you get a firewall.

"Quantum mechanics doesn't allow both to be there," Polchinski said. "If you lose the entanglement between the in-falling (Alice) and the outgoing (Bob) observers, it means you've put some kind of sharp kink into the quantum state right at the horizon. You've broken a bond, in some sense, and that broken bond requires energy. This tells us the firewall has to be there."

That consequence arises from the fact that entanglement between the area outside the event horizon and the Hawking radiation must increase as the black hole evaporates. When roughly half the mass has radiated away, the black hole is maximally entangled and essentially experiences a mid-life crisis. Preskill explained: "It's as if the singularity, which we expected to find deep inside the black hole, has crept right up to the event horizon when the black hole is old." And the result of this collision between the singularity and the event horizon is the dreaded firewall.

The mental image of a singularity migrating from deep within a black hole to the event horizon provoked at least one exasperated outburst during the Stanford workshop, a reaction Bousso finds understandable.[3] "We should be upset," he said. "This is a terrible blow to general relativity."

Yet for all his skepticism about firewalls, he is thrilled to be part of the debate. "This is probably the most exciting thing that's happened to me since I entered physics," he said. "It's certainly the nicest paradox that's come my way, and I'm excited to be working on it."

Alice's death by firewall seems destined to join the ranks of classic thought experiments in physics. The more physicists learn about quantum gravity, the more different it appears to be from our current picture of how the universe works, forcing them to sacrifice one cherished belief after another on the altar of scientific progress. Now they must choose to sacrifice either unitarity or No Drama, or undertake a radical modification of quantum field theory. Or maybe it's all just a horrible mistake. Any way you slice it, physicists are bound to learn something new.

WORMHOLES UNTANGLE A BLACK HOLE PARADOX

K. C. Cole

One hundred years after Albert Einstein developed his general theory of relativity, physicists are still stuck with perhaps the biggest incompatibility problem in the universe. The smoothly warped space-time landscape that Einstein described is like a painting by Salvador Dalí—seamless, unbroken, geometric. But the quantum particles that occupy this space are more like something from Georges Seurat: pointillist, discrete, described by probabilities. At their core, the two descriptions contradict each other. Yet a bold new strain of thinking suggests that quantum correlations between specks of impressionist paint actually create not just Dalí's landscape, but the canvases that both sit on, as well as the three-dimensional space around them. And Einstein, as he so often does, sits right in the center of it all, still turning things upside-down from beyond the grave.

Like initials carved in a tree, ER=EPR, as the new idea is known, is a shorthand that joins two ideas proposed by Einstein in 1935. One involved the paradox implied by what he called "spooky action at a distance" between quantum particles (the EPR paradox, named for its authors, Einstein, Boris Podolsky and Nathan Rosen). The other showed how two black holes could be connected through far reaches of space through "wormholes" (ER, for Einstein-Rosen bridges). At the time that Einstein put forth these ideas—and for most of the eight decades since—they were thought to be entirely unrelated.

But if ER=EPR is correct, the ideas aren't disconnected—they're two manifestations of the same thing. And this underlying connectedness would form the foundation of all space-time. Quantum entanglement—the action at a distance that so troubled Einstein—could be creating the "spatial connectivity" that "sews space together," according to Leonard Susskind, a physicist at Stanford University and one of the idea's main architects. Without these connections, all of space would "atomize," according to Juan Maldacena, a physicist at the Institute for Advanced Study in Princeton, New Jersey, who developed the idea together with Susskind. "In other words, the solid and reliable structure of space-time is due to the ghostly features of entanglement," he said. What's more, ER = EPR has the potential to address how gravity fits together with quantum mechanics.

Not everyone's buying it, of course (nor should they; the idea is in "its infancy," said Susskind). Joseph Polchinski, whose own stunning paradox about firewalls in the throats of black holes triggered the latest advances, was cautious, but intrigued, when asked about it in 2015. "I don't know where it's going," he said, "but it's a fun time right now."

THE BLACK HOLE WARS

The road that led to ER = EPR is a Möbius strip of tangled twists and turns that folds back on itself, like a drawing by M. C. Escher.

A fair place to start might be quantum entanglement. If two quantum particles are entangled, they become, in effect, two parts of a single unit. What happens to one entangled particle happens to the other, no matter how far apart they are.

Maldacena sometimes uses a pair of gloves as an analogy: If you come upon the right-handed glove, you instantaneously know the other is left-handed. There's nothing spooky about that. But in the quantum version, both gloves are actually left- and right-handed (and everything in between) up until the moment you observe them. Spookier still, the left-handed glove doesn't become left until you observe the right-handed one—at which moment both instantly gain a definite handedness.

Entanglement played a key role in Stephen Hawking's 1974 discovery that black holes could evaporate. This, too, involved entangled pairs of particles. Throughout space, short-lived "virtual" particles of matter and antimatter continually pop into and out of existence. Hawking realized that if one particle fell into a black hole and the other escaped, the hole would emit radiation, glowing like a dying ember. Given enough time, the hole would evaporate into nothing, raising the question of what happened to the information content of the stuff that fell into it.

But the rules of quantum mechanics forbid the complete destruction of information. (Hopelessly scrambling information is another story, which is why documents can be burned and hard drives smashed. There's nothing in the laws of physics that prevents the information lost in a book's smoke and ashes from being reconstructed, at least in principle.) So the question became: Would the information that originally went into the black hole just get scrambled? Or would it be truly lost? The arguments set off what Susskind called the "black hole wars," which have generated enough stories to fill many books. (Susskind's was subtitled "My Battle with Stephen Hawking to Make the World Safe for Quantum Mechanics.")

Eventually Susskind—in a discovery that shocked even him—realized (with Gerard 't Hooft) that all the information that fell down the hole was actually trapped on the black hole's two-dimensional event horizon, the surface that marks the point of no return. The horizon encoded everything inside, like a hologram. It was as if the bits needed to re-create your house and everything in it could fit on the walls. The information wasn't lost—it was scrambled and stored out of reach.

Susskind continued to work on the idea with Maldacena, whom Susskind calls "the master," and others. Holography began to be used not just to understand black holes, but any region of space that can be described by its boundary. Over the past decade or so, the seemingly crazy idea that space is a kind of hologram has become rather humdrum, a tool of modern physics used in everything from cosmology to condensed matter. "One of the things that happens to scientific ideas is they often go from wild conjecture to reasonable conjecture to working tools," Susskind said. "It's gotten routine."

Holography was concerned with what happens on boundaries, including black hole horizons. That left open the question of what goes on in the interiors, said Susskind, and answers to that "were all over the map." After all, since no information could ever escape from inside a black hole's horizon, the laws of physics prevented scientists from ever directly testing what was going on inside.

Then in 2012 Polchinski, along with Ahmed Almheiri, Donald Marolf and James Sully, all of them at the time at Santa Barbara, came up with an insight so startling it basically said to physicists: Hold everything. We know nothing.

The so-called AMPS paper (after its authors' initials) presented a doozy of an entanglement paradox—one so stark it implied that black holes might not, in effect, even have insides, for a "firewall" just inside the horizon would fry anyone or anything attempting to find out its secrets.[1]

The AMPS paper became a "real trigger," said Stephen Shenker, a physicist at Stanford, and "cast in sharp relief" just how much was not understood. Of course, physicists love such paradoxes, because they're fertile ground for discovery.

Both Susskind and Maldacena got on it immediately. They'd been thinking about entanglement and wormholes, and both were inspired by the work of Mark Van Raamsdonk, a physicist at the University of British Columbia in Vancouver, who had conducted a pivotal thought experiment suggesting that entanglement and space-time are intimately related.

"Then one day," said Susskind, "Juan sent me a very cryptic message that contained the equation ER=EPR. I instantly saw what he was getting at, and from there we went back and forth expanding the idea."

Their investigations, which they presented in a 2013 paper, "Cool Horizons for Entangled Black Holes," argued for a kind of entanglement they said the AMPS authors had overlooked—the one that "hooks space together," according to Susskind.[2] AMPS assumed that the parts of space inside and outside of the event horizon were independent. But Susskind and Maldacena suggest that, in fact, particles on either side of the border could be connected by a wormhole. The ER=EPR entanglement could "kind of get around the apparent paradox," said Van Raamsdonk. The paper contained a graphic that some refer to half-jokingly as the "octopus picture"—with multiple wormholes leading from the inside of a black hole to Hawking radiation on the outside.

In other words, there was no need for an entanglement that would create a kink in the smooth surface of the black hole's throat. The particles still inside the hole would be directly connected to particles that left long ago. No need to pass through the horizon, no need to pass Go. The particles on the inside and the far-out ones could be considered one and the same, Maldacena explained—like me, myself and I. The complex "octopus" wormhole would link the interior of the black hole directly to particles in the long-departed cloud of Hawking radiation.

HOLES IN THE WORMHOLE

No one is sure yet whether ER=EPR will solve the firewall problem. John Preskill reminded readers of *Quantum Frontiers*, the blog for Caltech's Institute for Quantum Information and Matter, that sometimes physicists rely on their "sense of smell" to sniff out which theories have promise. "At first whiff," he wrote, "ER=EPR may smell fresh and sweet, but it will have to ripen on the shelf for a while."

Whatever happens, the correspondence between entangled quantum particles and the geometry of smoothly warped space-time is a "big new insight," said Shenker. It's allowed him and his collaborator Douglas Stanford, a researcher at the Institute for Advanced Study, to tackle complex problems in quantum chaos through what Shenker calls "simple geometry that even I can understand."[3]

To be sure, ER=EPR does not yet apply to just any kind of space, or any kind of entanglement. It takes a special type of entanglement and a special type of wormhole. "Lenny and Juan are completely aware of this," said

Marolf, who co-authored a paper describing wormholes with more than two ends.[4] ER=EPR works in very specific situations, he said, but AMPS argues that the firewall presents a much broader challenge.

Like Polchinski and others, Marolf worried that ER=EPR modifies standard quantum mechanics. "A lot of people are really interested in the ER=EPR conjecture," said Marolf. "But there's a sense that no one but Lenny and Juan really understand what it is." Still, "it's an interesting time to be in the field."

HOW QUANTUM PAIRS STITCH SPACE-TIME

Jennifer Ouellette

B rian Swingle was a graduate student studying the physics of matter at the Massachusetts Institute of Technology when he decided to take a few classes in string theory to round out his education—"because, why not?" he recalled—although he initially paid little heed to the concepts he encountered in those classes. But as he delved deeper, he began to see unexpected similarities between his own work, in which he used so-called tensor networks to predict the properties of exotic materials, and string theory's approach to black-hole physics and quantum gravity. "I realized there was something profound going on," he said.

Tensors crop up all over physics—they're simply mathematical objects that can represent multiple numbers at the same time. For example, a velocity vector is a simple tensor: It captures values for both the speed and the direction of motion. More complicated tensors, linked together into networks, can be used to simplify calculations for complex systems made of many different interacting parts—including the intricate interactions of the vast numbers of subatomic particles that make up matter.

Swingle is one of a growing number of physicists who see the value in adapting tensor networks to cosmology. Among other benefits, it could help resolve an ongoing debate about the nature of space-time itself. According to John Preskill, many physicists have suspected a deep connection between quantum entanglement—the "spooky action at a distance" that so vexed Albert Einstein—and space-time geometry at the smallest scales since the physicist John Wheeler first described the latter as a bubbly, frothy foam six decades ago. "If you probe geometry at scales comparable to the Planck scale"—the shortest possible distance—"it looks less and less like space-time," said Preskill. "It's not really geometry anymore. It's something else, an emergent thing [that arises] from something more fundamental."

Physicists continue to wrestle with the knotty problem of what this more fundamental picture might be, but they strongly suspect that it is related to quantum information. "When we talk about information being encoded, [we mean that] we can split a system into parts, and there is some correlation among the parts so I can learn something about one part by observing another part," said Preskill. This is the essence of entanglement.

It is common to speak of a "fabric" of space-time, a metaphor that evokes the concept of weaving individual threads together to form a smooth, continuous whole. That thread is fundamentally quantum. "Entanglement is the fabric of space-time," said Swingle in 2015, then a researcher at Stanford University. "It's the thread that binds the system together, that makes the collective properties different from the individual properties. But to really see the interesting collective behavior, you need to understand how that entanglement is distributed."

Tensor networks provide a mathematical tool capable of doing just that. In this view, space-time arises out of a series of interlinked nodes in a complex network, with individual morsels of quantum information fitted together like Legos. Entanglement is the glue that holds the network together. If we want to understand space-time, we must first think geometrically about entanglement, since that is how information is encoded between the immense number of interacting nodes in the system.

MANY BODIES, ONE NETWORK

It is no easy feat to model a complex quantum system; even doing so for a classical system with more than two interacting parts poses a challenge. When Isaac Newton published his *Principia* in 1687, one of the many topics he examined became known as the "three-body problem." It is a relatively simple matter to calculate the movement of two objects, such as the Earth and the sun, taking into account the effects of their mutual gravitational attraction. However, adding a third body, like the moon, turns a relatively straightforward problem with an exact solution into one that is inherently chaotic, where long-term predictions require powerful computers to simulate an approximation of the system's evolution. In general, the more objects in the system, the more difficult the calculation, and that difficulty increases linearly, or nearly so—at least in classical physics.

Now imagine a quantum system with many billions of atoms, all of which interact with each other according to complicated quantum equations. At that scale, the difficulty appears to increase exponentially with the number of particles in the system, so a brute-force approach to calculation just won't work.

Consider a lump of gold. It is comprised of many billions of atoms, all of which interact with one another. From those interactions emerge the various classical properties of the metal, such as color, strength or conductivity. "Atoms are tiny little quantum mechanical things, and you put atoms together and new and wonderful things happen," said Swingle. But at this scale, the rules of quantum mechanics apply. Physicists need to precisely calculate the wave function of that lump of gold, which describes the state of the system. And that wave function is a many-headed hydra of exponential complexity.

Even if your lump of gold has just 100 atoms, each with a quantum "spin" that can be either up or down, the number of possible states totals 2^{100}, or a million trillion trillion. With every added atom the problem grows exponentially worse. (And worse still if you care to describe anything in addition to the atomic spins, which any realistic model would.) "If you take the entire visible universe and fill it up with our best storage material, the best hard drive money can buy, you could only store the state of about 300 spins," said Swingle. "So this information is there, but it's not all physical. No one has ever measured all these numbers."

Tensor networks enable physicists to compress all the information contained within the wave function and focus on just those properties physicists can measure in experiments: how much a given material bends light, for example, or how much it absorbs sound, or how well it conducts electricity. A tensor is a "black box" of sorts that takes in one collection of numbers and spits out a different one. So it is possible to plug in a simple wave function—such as that of many non interacting electrons, each in its lowest-energy state—and run tensors upon the system over and over, until the process produces a wave function for a large, complicated system, like the billions of interacting atoms in a lump of gold. The result is a straightforward diagram that represents this complicated lump of gold, an innovation much like the development of Feynman diagrams in the mid-20th century, which simplified how physicists represent particle interactions. A tensor network has a geometry, just like space-time.

The key to achieving this simplification is a principle called "locality." Any given electron only interacts with its nearest neighboring electrons. Entangling each of many electrons with its neighbors produces a series of "nodes" in the network. Those nodes are the tensors, and entanglement links them together. All those interconnected nodes make up the network. A complex calculation thus becomes easier to visualize. Sometimes it even reduces to a much simpler counting problem.

There are many different types of tensor networks, but among the most useful is the one known by the acronym MERA (multiscale entanglement renormalization ansatz). Here's how it works in principle: Imagine a one-dimensional line of electrons. Replace the eight individual electrons—designated A, B, C, D, E, F, G and H—with fundamental units of quantum information (qubits), and entangle them with their nearest neighbors to form links. A entangles with B, C entangles with D, E entangles with F, and G entangles with H. This produces a higher level in the network. Now entangle AB with CD, and EF with GH, to get the next level in the network. Finally, ABCD entangles with EFGH to form the highest layer. "In a way, we could say that one uses entanglement to build up the many-body wave function," Román Orús, a physicist at Johannes Gutenberg University in Germany, observed in a 2014 paper.[1]

Why are some physicists so excited about the potential for tensor networks—especially MERA—to illuminate a path to quantum gravity? Because the networks demonstrate how a single geometric structure can emerge from complicated interactions between many objects. And Swingle (among others) hopes to make use of this emergent geometry by showing how it can explain the mechanism by which a smooth, continuous space-time can emerge from discrete bits of quantum information.

SPACE-TIME'S BOUNDARIES

Condensed-matter physicists inadvertently found an emergent extra dimension when they developed tensor networks: the technique yields a two-dimensional system out of one dimension. Meanwhile, gravity theorists were subtracting a dimension—going from three to two—with the development of what's known as the holographic principle. The two concepts might connect to form a more sophisticated understanding of space-time.

In the 1970s, a physicist named Jacob Bekenstein showed that the information about a black hole's interior is encoded in its two-dimensional surface area (the "boundary") rather than within its three-dimensional volume (the "bulk"). Twenty years later, Leonard Susskind and Gerard 't Hooft extended this notion to the entire universe, likening it to a hologram: Our three-dimensional universe in all its glory emerges from a two-dimensional "source code." In 1997, Juan Maldacena found a concrete example of holography in action, demonstrating that a toy model describing a flat space without gravity is equivalent to a description of a saddle-shaped space with gravity. This connection is what physicists call a "duality."

Mark Van Raamsdonk likens the holographic concept to a two-dimensional computer chip that contains the code for creating the three-dimensional virtual world of a video game. We live within that 3-D game space. In one sense, our space is illusory, an ephemeral image projected into thin air. But as Van Raamsdonk emphasizes, "There's still an actual physical thing in your computer that stores all the information."

The idea has gained broad acceptance among theoretical physicists, but they still grapple with the problem of precisely how a lower dimension would store information about the geometry of space-time. The sticking point is that our metaphorical memory chip has to be a kind of quantum computer, where the traditional zeros and ones used to encode information are replaced with qubits capable of being zeros, ones and everything in between simultaneously. Those qubits must be connected via entanglement—whereby the state of one qubit is determined by the state of its neighbor—before any realistic 3-D world can be encoded.

Similarly, entanglement seems to be fundamental to the existence of space-time. This was the conclusion reached by a pair of postdocs in 2006: Shinsei Ryu (now at the University of Chicago) and Tadashi Takayanagi (now at Kyoto University), who shared the 2015 New Horizons in Physics prize for this work.[2] "The idea was that the way that [the geometry of] space-time is encoded has a lot to do with how the different parts of this memory chip are entangled with each other," Van Raamsdonk explained.

Inspired by their work, as well as by a subsequent paper of Maldacena's, in 2010 Van Raamsdonk proposed a thought experiment to demonstrate the critical role of entanglement in the formation of space-time, pondering what would happen if one cut the memory chip in two and then removed the entanglement between qubits in opposite halves. He found that space-time begins to tear itself apart, in much the same way that stretching a wad of gum by both ends yields a pinched-looking point in the center as the two halves move farther apart. Continuing to split that memory chip into smaller and smaller pieces unravels space-time until only tiny individual fragments remain that have no connection to one another. "If you take away the entanglement, your space-time just falls apart," said Van Raamsdonk. Similarly, "if you wanted to build up a space-time, you'd want to start entangling [qubits] together in particular ways."

Combine those insights with Swingle's work connecting the entangled structure of space-time and the holographic principle to tensor networks, and another crucial piece of the puzzle snaps into place. Curved space-times emerge quite naturally from entanglement in tensor networks via holography.[3] "Space-time is a geometrical representation of this quantum information," said Van Raamsdonk.

And what does that geometry look like? In the case of Maldacena's saddle-shaped space-time, it looks like one of M. C. Escher's *Circle Limit* figures from the late 1950s and early 1960s. Escher had long been interested in order and symmetry, incorporating those mathematical concepts into his art ever since 1936 when he visited the Alhambra in Spain, where he found inspiration in the repeating tiling patterns typical of Moorish architecture, known as tessellation.

His *Circle Limit* woodcuts are illustrations of hyperbolic geometries: negatively curved spaces represented in two dimensions as a distorted disk, much the way flattening a globe into a two-dimensional map of the Earth distorts the continents. For instance, *Circle Limit IV (Heaven and Hell)* features many repeating figures of angels and demons. In a true hyperbolic space, all the figures would be the same size, but in Escher's two-dimensional representation, those near the edge appear smaller and more pinched than the figures in the center. A diagram of a tensor network also bears a striking resemblance to the *Circle Limit* series, a visual manifestation of the deep connection Swingle noticed when he took that fateful string theory class.

To date, tensor analysis has been limited to models of space-time, like Maldacena's, that don't describe the universe we inhabit—a non-saddle-shaped universe whose expansion is accelerating. Physicists can only translate between dual models in a few special cases. Ideally, they would like to have a universal dictionary. And they would like to be able to derive that dictionary directly, rather than make close approximations. "We're in a funny situation with these dualities, because everyone seems to agree that it's important, but nobody knows how to derive them," said Preskill. "Maybe the tensor-network approach will make it possible to go further. I think it would be a sign of progress if we can say—even with just a toy model—'Aha! Here is the derivation of the dictionary!' That would be a strong hint that we are onto something."

Swingle and Van Raamsdonk have collaborated to move their respective work in this area beyond a static picture of space-time to explore its dynamics: how space-time changes over time, and how it curves in response to these changes. They have managed to derive Einstein's equations, specifically the equivalence principle—evidence that the dynamics of space-time, as well as its geometry, emerge from entangled qubits. It is a promising start.

"'What is space-time?' sounds like a completely philosophical question," Van Raamsdonk said. "To actually have some answer to that, one that is concrete and allows you to calculate space-time, is kind of amazing."

IN A MULTIVERSE, WHAT ARE THE ODDS?

Natalie Wolchover

f modern physics is to be believed, we shouldn't be here. The meager dose of energy infusing empty space, which at higher levels would rip the cosmos apart, is a trillion trillion trillion trillion trillion trillion trillion trillion trillion trillion times tinier than theory predicts. And the minuscule mass of the Higgs boson, whose relative smallness allows big structures such as galaxies and humans to form, falls roughly 100 quadrillion times short of expectations. Dialing up either of these constants even a little would render the universe unlivable.

To account for our incredible luck, leading cosmologists like Alan Guth envision our universe as one of countless bubbles in an eternally frothing sea. This infinite "multiverse" would contain universes with constants tuned to any and all possible values, including some outliers, like ours, that have just the right properties to support life. In this scenario, our good luck is inevitable: A peculiar, life-friendly bubble is all we could expect to observe.

The problem remains how to test the hypothesis. Proponents of the multiverse idea must show that, among the rare universes that support life, ours is statistically typical. The exact dose of vacuum energy, the precise mass of our underweight Higgs boson, and other anomalies must have high odds within the subset of habitable universes. If the properties of this universe still seem atypical even in the habitable subset, then the multiverse explanation fails.

But infinity sabotages statistical analysis. In an eternally inflating multiverse, where any bubble that can form does so infinitely many times, how do you measure "typical"?

Guth, a professor of physics at the Massachusetts Institute of Technology, resorts to freaks of nature to pose this "measure problem." "In a single universe, cows born with two heads are rarer than cows born with one head," he said. But in an infinitely branching multiverse, "there are an infinite number of one-headed cows and an infinite number of two-headed cows. What happens to the ratio?"

For years, the inability to calculate ratios of infinite quantities has prevented the multiverse hypothesis from making testable predictions about the properties of this universe. For the hypothesis to mature into a full-fledged theory of physics, the two-headed-cow question demands an answer.

ETERNAL INFLATION

As a junior researcher trying to explain the smoothness and flatness of the universe, Guth proposed in 1980 that a split second of exponential growth may have occurred at the start of the Big Bang.[1] This would have ironed out any spatial variations as if they were wrinkles on the surface of an inflating balloon. The inflation hypothesis, though it is still being tested, gels with all available astrophysical data and is widely accepted by physicists.

In the years that followed, Andrei Linde, now of Stanford University, Guth and other cosmologists reasoned that inflation would almost inevitably beget an infinite number of universes. "Once inflation starts, it never stops completely," Guth explained. In a region where it does stop—through a kind of decay that settles it into a stable state—space and time gently swell into a universe like ours. Everywhere else, space-time continues to expand exponentially, bubbling forever.

Each disconnected space-time bubble grows under the influence of different initial conditions tied to decays of varying amounts of energy. Some bubbles expand and then contract, while others spawn endless streams of daughter universes. The scientists presumed that the eternally inflating multiverse would everywhere obey the conservation of energy, the speed of light, thermodynamics, general relativity and quantum mechanics. But the values of the constants coordinated by these laws were likely to vary randomly from bubble to bubble.

Paul Steinhardt, a theoretical physicist at Princeton University and one of the early contributors to the theory of eternal inflation, saw the multiverse as a "fatal flaw" in the reasoning he had helped advance, and he remains stridently anti-multiverse today. "Our universe has a simple, natural structure," he said in 2014. "The multiverse idea is baroque, unnatural, untestable and, in the end, dangerous to science and society."

Steinhardt and other critics believe the multiverse hypothesis leads science away from uniquely explaining the properties of nature. When deep questions about matter, space and time have been elegantly answered over the past century through ever more powerful theories, deeming the universe's remaining unexplained properties "random" feels, to them, like giving up. On the other hand, randomness has sometimes been the answer to scientific questions, as when early astronomers searched in vain for order in the solar system's haphazard planetary orbits. As inflationary cosmology gains acceptance, more physicists are conceding that a multiverse of random universes might exist, just as there is a cosmos full of star systems arranged by chance and chaos.

"When I heard about eternal inflation in 1986, it made me sick to my stomach," said John Donoghue, a physicist at the University of Massachusetts, Amherst. "But when I thought about it more, it made sense."

ONE FOR THE MULTIVERSE

The multiverse hypothesis gained considerable traction in 1987, when the Nobel laureate Steven Weinberg used it to predict the infinitesimal amount of energy infusing the vacuum of empty space, a number known as the cosmological constant, denoted by the Greek letter Λ (lambda). Vacuum energy is gravitationally repulsive, meaning it causes space-time to stretch apart. Consequently, a universe with a positive value for Λ expands—faster and faster, in fact, as the amount of empty space grows—toward a future as a matter-free void. Universes with negative Λ eventually contract in a "big crunch."

Physicists had not yet measured the value of Λ in our universe in 1987, but the relatively sedate rate of cosmic expansion indicated that its value was close to zero. This flew in the face of quantum mechanical calculations suggesting Λ should be enormous, implying a density of vacuum energy so large it would tear atoms apart. Somehow, it seemed our universe was greatly diluted.

Weinberg turned to a concept called anthropic selection in response to "the continued failure to find a microscopic explanation of the smallness of the cosmological constant," as he wrote in *Physical Review Letters (PRL)*. He posited that life forms, from which observers of universes are drawn, require the existence of galaxies. The only values of Λ that can be observed are therefore those that allow the universe to expand slowly enough for matter to clump together into galaxies. In his *PRL* paper, Weinberg reported the maximum possible value of Λ in a universe that has galaxies.[2] It was a multiverse-generated prediction of the most likely density of vacuum energy to be observed, given that observers must exist to observe it.

A decade later, astronomers discovered that the expansion of the cosmos was accelerating at a rate that pegged Λ at 10^{-123} (in units of "Planck energy density"). A value of exactly zero might have implied an unknown symmetry in the laws of quantum mechanics—an explanation without a multiverse. But this absurdly tiny value of the cosmological constant appeared random. And it fell strikingly close to Weinberg's prediction.

"It was a tremendous success, and very influential," said Matthew Kleban, a multiverse theorist at New York University. The prediction seemed to show that the multiverse could have explanatory power after all.

Close on the heels of Weinberg's success, Donoghue and colleagues used the same anthropic approach to calculate the range of possible values for the mass of the Higgs boson. The Higgs doles out mass to other elementary particles, and these interactions dial its mass up or down in a feedback effect. This feedback would be expected to yield a mass for the Higgs that is far larger than its observed value, making its mass appear to have been reduced by accidental cancellations between the effects of all the individual particles. Donoghue's group argued that this accidentally tiny Higgs was to be expected, given anthropic selection: If the Higgs boson were just five times heavier, then complex, life-engendering elements like carbon could not arise.[3] Thus, a universe with much heavier Higgs particles could never be observed.

Until recently, the leading explanation for the smallness of the Higgs mass was a theory called supersymmetry, but the simplest versions of the theory have failed extensive tests at the Large Hadron Collider near Geneva. Although new alternatives have been proposed, many particle physicists who considered the multiverse unscientific just a few years ago are now grudgingly opening up to the idea. "I wish it would go away," said Nathan Seiberg, a professor of physics at the Institute for Advanced Study, who contributed to supersymmetry in the 1980s. "But you have to face the facts."

However, even as the impetus for a predictive multiverse theory has increased, researchers have realized that the predictions by Weinberg and others were too naive. Weinberg estimated the largest Λ compatible with the formation of galaxies, but that was before astronomers discovered mini "dwarf galaxies" that could form in universes in which Λ is 1,000 times larger.[4] These more prevalent universes can also contain observers, making our universe seem atypical among observable universes. On the other hand, dwarf galaxies presumably contain fewer observers than full-size ones, and universes with only dwarf galaxies would therefore have lower odds of being observed.

Researchers realized it wasn't enough to differentiate between observable and unobservable bubbles. To accurately predict the expected properties of our universe, they needed to weight the likelihood of observing certain bubbles according to the number of observers they contained. Enter the measure problem.

MEASURING THE MULTIVERSE

Guth and other scientists sought a measure to gauge the odds of observing different kinds of universes. This would allow them to make predictions about the assortment of fundamental constants in this universe, all of which should have reasonably high odds of being observed. The scientists' early attempts involved constructing mathematical models of eternal inflation and calculating the statistical distribution of observable bubbles based on how many of each type arose in a given time interval. But with time serving as the measure, the final tally of universes at the end depended on how the scientists defined time in the first place.

"People were getting wildly different answers depending on which random cutoff rule they chose," said Raphael Bousso of the University of California, Berkeley.

Alex Vilenkin, director of the Institute of Cosmology at Tufts University in Medford, Massachusetts, has proposed and discarded several multiverse measures during the last two decades, looking for one that would transcend his arbitrary assumptions. In 2012, he and Jaume Garriga of the University of Barcelona in Spain proposed a measure in the form of an immortal "watcher" who soars through the multiverse counting events, such as the number of observers.[5] The frequencies of events are then converted to probabilities, thus solving the measure problem. But the proposal assumes the impossible up front: The watcher miraculously survives crunching bubbles, like an avatar in a video game dying and bouncing back to life.

In 2011, Guth and Vitaly Vanchurin, now of the University of Minnesota Duluth, imagined a finite "sample space," a randomly selected slice of space-time within the infinite multiverse.[6] As the sample space expands, approaching but never reaching infinite size, it cuts through bubble universes encountering events, such as proton formations, star formations or intergalactic wars. The events are logged in a hypothetical databank until the sampling ends. The relative frequency of different events translates into probabilities and thus provides a predictive power. "Anything that can happen will happen, but not with equal probability," Guth said.

Still, beyond the strangeness of immortal watchers and imaginary databanks, both of these approaches necessitate arbitrary choices about which events should serve as proxies for life, and thus for observations of universes to be counted and converted into probabilities. Protons seem necessary for life; space wars do not—but do observers require stars, or is this too limited a concept of life? With either measure, choices can be made so that the odds stack in favor of our inhabiting a universe like ours. The degree of speculation raises doubts.

THE CAUSAL DIAMOND

Bousso first encountered the measure problem in the 1990s as a graduate student working with Stephen Hawking, the doyen of black hole physics. Black holes prove there is no such thing as an omniscient measurer, because someone inside a black hole's "event horizon," beyond which no light can escape, has access to different information and events from someone outside, and vice versa. Bousso and other black hole specialists came to think such a rule "must be more general," he said, precluding solutions to the measure problem along the lines of the immortal watcher. "Physics is universal, so we've got to formulate what an observer can, in principle, measure."

This insight led Bousso to develop a multiverse measure that removes infinity from the equation altogether.[7] Instead of looking at all of spacetime, he homes in on a finite patch of the multiverse called a "causal diamond," representing the largest swath accessible to a single observer traveling from the beginning of time to the end of time. The finite boundaries of a causal diamond are formed by the intersection of two cones of light, like the dispersing rays from a pair of flashlights pointed toward each other in the dark. One cone points outward from the moment matter was created after a Big Bang—the earliest conceivable birth of an observer—and the other aims backward from the farthest reach of our future horizon, the moment when the causal diamond becomes an empty, timeless void and the observer can no longer access information linking cause to effect.

Bousso is not interested in what goes on outside the causal diamond, where infinitely variable, endlessly recursive events are unknowable, in the same way that information about what goes on outside a black hole cannot be accessed by the poor soul trapped inside. If one accepts that the finite diamond, "being all anyone can ever measure, is also all there is," Bousso said, "then there is indeed no longer a measure problem."

In 2006, Bousso realized that his causal-diamond measure lent itself to an evenhanded way of predicting the expected value of the cosmological constant. Causal diamonds with smaller values of Λ would produce more entropy—a quantity related to disorder, or degradation of energy—and Bousso postulated that entropy could serve as a proxy for complexity and thus for the presence of observers. Unlike other ways of counting observers, entropy can be calculated using trusted thermodynamic equations. With this approach, Bousso said, "comparing universes is no more exotic than comparing pools of water to roomfuls of air."

Using astrophysical data, Bousso and his collaborators Roni Harnik, Graham Kribs and Gilad Perez calculated the overall rate of entropy production in our universe, which primarily comes from light scattering off cosmic dust.[8] The calculation predicted a statistical range of expected values of Λ. The known value, 10^{-123}, rests just left of the median. "We honestly didn't see it coming," Bousso said. "It's really nice, because the prediction is very robust."

MAKING PREDICTIONS

Bousso and his collaborators' causal-diamond measure has now racked up a number of successes. It offers a solution to a mystery of cosmology called the "why now?" problem, which asks why we happen to live at a time when the effects of matter and vacuum energy are comparable, so that the expansion of the universe recently switched from slowing down (signifying a matter-dominated epoch) to speeding up (a vacuum energy-dominated epoch). Bousso's theory suggests it is only natural that we find ourselves at this juncture. The most entropy is produced, and therefore the most observers exist, when universes contain equal parts vacuum energy and matter.

In 2010 Harnik and Bousso used their idea to explain the flatness of the universe and the amount of infrared radiation emitted by cosmic dust. In 2013, Bousso and his Berkeley colleague Lawrence Hall reported that observers made of protons and neutrons, like us, will live in universes where the amount of ordinary matter and dark matter are comparable, as is the case here.[9]

"Right now the causal patch looks really good," Bousso said. "A lot of things work out unexpectedly well, and I do not know of other measures that come anywhere close to reproducing these successes or featuring comparable successes."

The causal-diamond measure falls short in a few ways, however. It does not gauge the probabilities of universes with negative values of the cosmological constant. And its predictions depend sensitively on assumptions about the early universe, at the inception of the future-pointing light cone.

But researchers in the field recognize its promise. By sidestepping the infinities underlying the measure problem, the causal diamond "is an oasis of finitude into which we can sink our teeth," said Andreas Albrecht, a theoretical physicist at the University of California, Davis, and one of the early architects of inflation.

Kleban, who like Bousso began his career as a black hole specialist, said the idea of a causal patch such as an entropy-producing diamond is "bound to be an ingredient of the final solution to the measure problem." He, Guth, Vilenkin and many other physicists consider it a powerful and compelling approach, but they continue to work on their own measures of the multiverse. Few consider the problem to be solved.

Every measure involves many assumptions, beyond merely that the multiverse exists. For example, predictions of the expected range of constants like Λ and the Higgs mass always speculate that bubbles tend to have larger constants. Clearly, this is a work in progress.

"The multiverse is regarded either as an open question or off the wall," Guth said. "But ultimately, if the multiverse does become a standard part of science, it will be on the basis that it's the most plausible explanation of the fine-tunings that we see in nature."

Perhaps these multiverse theorists have chosen a Sisyphean task. Perhaps they will never settle the two-headed-cow question. Some researchers are taking a different route to testing the multiverse. Rather than rifle through the infinite possibilities of the equations, they are scanning the finite sky for the ultimate Hail Mary pass—the faint tremor from an ancient bubble collision.

MULTIVERSE COLLISIONS MAY DOT THE SKY

Jennifer Ouellette

L ike many of her colleagues, Hiranya Peiris, a cosmologist at University College London, once largely dismissed the notion that our universe might be only one of many in a vast multiverse. It was scientifically intriguing, she thought, but also fundamentally untestable. She preferred to focus her research on more concrete questions, like how galaxies evolve.

Then one summer at the Aspen Center for Physics, Peiris found herself chatting with the Perimeter Institute's Matt Johnson, who mentioned his interest in developing tools to study the idea. He suggested that they collaborate.

At first, Peiris was skeptical. "I think as an observer that any theory, however interesting and elegant, is seriously lacking if it doesn't have testable consequences," she said. But Johnson convinced her that there might be a way to test the concept. If the universe that we inhabit had long ago collided with another universe, the crash would have left an imprint on the cosmic microwave background (CMB), the faint afterglow from the Big Bang. And if physicists could detect such a signature, it would provide a window into the multiverse.

Erick Weinberg, a physicist at Columbia University, explains this multiverse by comparing it to a boiling cauldron, with the bubbles representing individual universes—isolated pockets of space-time. As the pot boils, the bubbles expand and sometimes collide. A similar process may have occurred in the first moments of the cosmos.

In the years since their initial meeting, Peiris and Johnson have studied how a collision with another universe in the earliest moments of time would have sent something similar to a shock wave across our universe. They think they may be able to find evidence of such a collision in data from the Planck space telescope, which maps the CMB.

The project might not work, Peiris concedes. It requires not only that we live in a multiverse but also that our universe collided with another in our primal cosmic history. But if physicists succeed, they will have the first improbable evidence of a cosmos beyond our own.

WHEN BUBBLES COLLIDE

Multiverse theories were once relegated to science fiction or crackpot territory. "It sounds like you've gone to crazy land," said Johnson, who holds joint appointments at the Perimeter Institute of Theoretical Physics and York University. But scientists have come up with many versions of what a multiverse might be, some less crazy than others.

The multiverse that Peiris and her colleagues are interested in is not the controversial "many worlds" hypothesis that was first proposed in the 1950s and holds that every quantum event spawns a separate universe. Nor is this concept of a multiverse related to the popular science-fiction trope of parallel worlds, new universes that pinch off from our space-time and become separate realms. Rather, this version arises as a consequence of inflation, a widely accepted theory of the universe's first moments.

Inflation holds that our universe experienced a sudden burst of rapid expansion an instant after the Big Bang, blowing up from a infinitesimally small speck to one spanning a quarter of a billion light-years in mere fractions of a second.

Yet inflation, once started, tends to never completely stop. According to the theory, once the universe starts expanding, it will end in some places, creating regions like the universe we see all around us today. But elsewhere inflation will simply keep on going eternally into the future.

This feature has led cosmologists to contemplate a scenario called eternal inflation. In this picture, individual regions of space stop inflating and become "bubble universes" like the one in which we live. But on larger scales, exponential expansion continues forever, and new bubble universes are continually being created. Each bubble is deemed a universe in its own right, despite being part of the same space-time, because an observer could not travel from one bubble to the next without moving faster than the speed of light. And each bubble may have its own distinct laws of physics. "If you buy eternal inflation, it predicts a multiverse," Peiris said.

In 2012, Peiris and Johnson teamed up with Anthony Aguirre and Max Wainwright to build a simulated multiverse with only two bubbles. They studied what happened after the bubbles collided to determine what an observer would see. The team concluded that a collision of two bubble universes would appear to us as a disk on the CMB with a distinctive temperature profile.

To guard against human error—we tend to see the patterns we want to see—they devised a set of algorithms to automatically search for these disks in data from the Wilkinson Microwave Anisotropy Probe (WMAP), a space-based observatory.[1] The program identified four potential regions with temperature fluctuations consistent with what could be a signature of a bubble collision. Later, after improving their theoretical predictions, they found that a stronger test is likely to come from CMB polarization data, not temperature fluctuations.[2] As new data from the Planck satellite becomes available, researchers should be able to improve on their analyses.

Yet detecting convincing signatures of the multiverse is tricky. Simply knowing what an encounter might look like requires a thorough understanding of the dynamics of bubble collisions—something quite difficult to model on a computer, given the complexity of such interactions.

When tackling a new problem, physicists typically find a good model that they already understand and adapt it by making minor tweaks they call "perturbations." For instance, to model the trajectory of a satellite in space, a physicist might use the classical laws of motion outlined by Isaac Newton in the 17th century and then make small refinements by calculating the effects of other factors that might influence its motion, such as pressure from the solar wind. For simple systems, there should be only small discrepancies from the unperturbed model. Try to calculate the airflow patterns of a complex system like a tornado, however, and those approximations break down. Perturbations introduce sudden, very large changes to the original system instead of smaller, predictable refinements.

Modeling bubble collisions during the inflationary period of the early universe is akin to modeling a tornado. By its very nature, inflation stretches out space-time at an exponential rate—precisely the kind of large jumps in values that make calculating the dynamics so challenging.

"Imagine you start with a grid, but within an instant, the grid has expanded to a massive size," Peiris said. With her collaborators, she has used techniques like adaptive mesh refinement—an iterative process of winnowing out the most relevant details in such a grid at increasingly finer scales—in her simulations of inflation to deal with the complexity. Eugene Lim, a physicist at King's College London, has found that an unusual type of traveling wave might help simplify matters even further.

WAVES OF TRANSLATION

In August 1834, a Scottish engineer named John Scott Russell was conducting experiments along Union Canal with an eye toward improving the efficiency of the canal boats. One boat being drawn by a team of horses stopped suddenly, and Russell noted a solitary wave in the water that kept rolling forward at a constant speed without losing its shape. The behavior was unlike typical waves, which tend to flatten out or rise to a peak and topple quickly. Intrigued, Russell tracked the wave on horseback for a couple of miles before it finally dissipated in the channel waters. This was the first recorded observation of a soliton.

Russell was so intrigued by the indomitable wave that he built a 30-foot wave tank in his garden to further study the phenomenon, noting key characteristics of what he called "the wave of translation." Such a wave could maintain size, shape and speed over longer distances than usual. The speed depended on the wave's size, and the width depended on the depth of the water. And if a large solitary wave overtook a smaller one, the larger, faster wave would just pass right through.

Russell's observations were largely dismissed by his peers because his findings seemed to contradict what was known about water wave physics at the time. It wasn't until the mid-1960s that such waves were dubbed solitons and physicists realized their usefulness in modeling problems in diverse areas such as fiber optics, biological proteins and DNA. Solitons also turn up in certain configurations of quantum field theory. Poke a quantum field and you will create an oscillation that usually dissipates outward, but configure things in just the right way and that oscillation will maintain its shape—just like Russell's wave of translation.

Because solitons are so stable, Lim believes they could work as a simplified toy model for the dynamics of bubble collisions in the multiverse, providing physicists with better predictions of what kinds of signatures might show up in the CMB. If his hunch is right, the expanding walls of our bubble universe are much like solitons.

However, while it is a relatively straightforward matter to model a solitary standing wave, the dynamics become vastly more complicated and difficult to calculate when solitons collide and interact, forcing physicists to rely on computer simulations instead. In the past, researchers have used a particular class of soliton with an exact mathematical solution and tweaked that model to suit their purposes. But this approach only works if the target system under study is already quite similar to the toy model; otherwise the changes are too large to calculate.

To get around that hurdle, Lim devised a neat trick based on a quirky feature of soliton collisions. When imagining two objects colliding, we naturally assume that the faster they are moving, the greater the impact and the more complicated the dynamics. Two cars ramming each other at high speeds, for instance, will produce scattered debris, heat, noise and other effects. The same is true for colliding solitons—at least initially. Collide two solitons very slowly, and there will be very little interaction, according to Lim. As the speed increases, the solitons interact more strongly.

But Lim found that as the speed continues to increase, the pattern eventually reverses: The soliton interaction begins to decrease. By the time they are traveling at the speed of light, there is no interaction at all. "They just fly right past each other," Lim said. "The faster you collide two solitons, the simpler they become." The lack of interactions makes it easier to model the dynamics of colliding solitons, as well as colliding bubble universes with solitons as their "edges," since the systems are roughly similar.[3]

According to Johnson, Lim has uncovered a very simple rule that can be applied broadly: Multiverse interactions are weak during high-speed collisions, making it easier to simulate the dynamics of those encounters. One can simply create a new model of the multiverse, use solitons as a tool to map the new model's expected signatures onto cosmic microwave data, and rule out any theories that don't match what researchers see. This process would help physicists identify the most viable models for the multiverse, which—while still speculative—would be consistent both with the latest observational data and with inflationary theory.

THE MULTIVERSE'S CASE FOR STRING THEORY

One reason that more physicists are taking the idea of the multiverse seriously is that certain such models could help resolve a significant challenge in string theory. One of the goals of string theory has been to unify quantum mechanics and general relativity, two separate "rule books" in physics that govern very different size scales, into a single, simple solution.

But around 15 years ago, "the dream of string theory kind of exploded," Johnson said—and not in a good way. Researchers began to realize that string theory doesn't provide a unique solution. Instead, it "gives you the theory of a vast number of worlds," Weinberg said. A common estimate—one that Weinberg thinks is conservative—is 10^{500} possibilities. This panoply of worlds implies that string theory can predict every possible outcome.

The multiverse would provide a possible means of incorporating all the different worlds predicted by string theory. Each version could be realized in its own bubble universe. "Everything depends on which part of the universe you live in," Lim said.

Peiris acknowledges that this argument has its critics. "It can predict anything, and therefore it's not valid," Peiris said of the reasoning typically used to dismiss the notion of a multiverse as a tautology, rather than a true scientific theory. "But I think that's the wrong way to think about it." The theory of evolution, Peiris argues, also resembles a tautology in certain respects—"an organism exists because it survived"—yet it holds tremendous explanatory power. It is a simple model that requires little initial input to produce the vast diversity of species we see today.

A multiverse model tied to eternal inflation could have the same kind of explanatory power. In this case, the bubble universes function much like speciation. Those universes that happen to have the right laws of physics will eventually "succeed"—that is, they will become home to conscious observers like ourselves. If our universe is one of many in a much larger multiverse, our existence seems less unlikely.

UNCERTAIN SIGNALS

Ultimately, however, Peiris' initial objection still stands: Without some means of gathering experimental evidence, the multiverse hypothesis will be untestable by definition. As such, it will lurk on the fringes of respectable physics—hence the strong interest in detecting bubble collision signatures in the CMB.

Of course, "just because these bubble collisions can leave a signature doesn't mean they do leave a signature," Peiris emphasized. "We need nature to be kind to us." An observable signal could be a rare find, given how quickly space expanded during inflation. The collisions may not have been rare, but subsequent inflation "tends to dilute away the effects of the collision just like it dilutes away all other prior 'structure' in the early universe, leaving you with a small chance of seeing a signal in the CMB sky," Peiris said.

"My own feeling is you need to adjust the numbers rather finely to get it to work," Weinberg said. The rate of formation of the bubble universes is key. If they had formed slowly, collisions would not have been possible because space would have expanded and driven the bubbles apart long before any collision could take place. Alternatively, if the bubbles had formed too quickly, they would have merged before space could expand sufficiently to form disconnected pockets. Somewhere in between is the Goldilocks rate, the "just right" rate at which the bubbles would have had to form for a collision to be possible.

Researchers also worry about finding a false positive. Even if such a collision did happen and evidence was imprinted on the CMB, spotting the tell-tale pattern would not necessarily constitute evidence of a multiverse. "You can get an effect and say it will be consistent with the calculated predictions for these [bubble] collisions," Weinberg said. "But it might well be consistent with lots of other things." For instance, a distorted CMB might be evidence of theoretical entities called cosmic strings. These are like the cracks that form in the ice when a lake freezes over, except here the ice is the fabric of space-time. Magnetic monopoles are another hypothetical defect that could affect the CMB, as could knots or twists in space-time called textures.

Weinberg isn't sure it would even be possible to tell the difference between these possibilities, especially because many models of eternal inflation exist. Without knowing the precise details of the theory, trying to make a positive identification of the multiverse would be like trying to distinguish between the composition of two meteorites that hit the roof of a house solely by the sound of the impacts, without knowing how the house is constructed and with what materials.

Should a signature for a bubble collision be confirmed, Peiris doesn't see a way to study another bubble universe any further because by now it would be entirely out of causal contact with ours. But it would be a stunning validation that the notion of a multiverse deserves a seat at the testable physics table.

And should that signal turn out to be evidence for cosmic strings or magnetic monopoles instead, it would still constitute exciting new physics at the frontier of cosmology. In that respect, "the cosmic microwave background radiation is the underpinning of modern cosmology," Peiris said. "It's the gift that keeps on giving."

HOW FEYNMAN DIAGRAMS ALMOST SAVED SPACE

Frank Wilczek

R ichard Feynman looked tired when he wandered into my office. It was the end of a long, exhausting day in Santa Barbara, sometime around 1982. Events had included a seminar that was also a performance, lunchtime grilling by eager postdocs and lively discussions with senior researchers. The life of a celebrated physicist is always intense. But our visitor still wanted to talk physics. We had a couple of hours to fill before dinner.

I described to Feynman what I thought were exciting if speculative new ideas such as fractional spin and anyons. Feynman was unimpressed, saying: "Wilczek, you should work on something real." (Anyons are real, but that's a topic for another time.[1])

Looking to break the awkward silence that followed, I asked Feynman the most disturbing question in physics, then as now: "There's something else I've been thinking a lot about: Why doesn't empty space weigh anything?"

Feynman, normally as quick and lively as they come, went silent. It was the only time I've ever seen him look wistful. Finally he said dreamily, "I once thought I had that one figured out. It was beautiful." And then, excited, he began an explanation that crescendoed in a near shout: "The reason space doesn't weigh anything, I thought, is because *there's nothing there!*"

To appreciate that surreal monologue, you need to know some backstory. It involves the distinction between vacuum and void.

I.

Vacuum, in modern usage, is what you get when you remove everything that you can, whether practically or in principle. We say a region of space "realizes vacuum" if it is free of all the different kinds of particles and radiation we know about (including, for this purpose, dark matter—which we know about in a general way, though not in detail). Alternatively, vacuum is the state of minimum energy.

Intergalactic space is a good approximation to a vacuum.

Void, on the other hand, is a theoretical idealization. It means nothingness: space without independent properties, whose only role, we might say, is to keep everything from happening in the same place. Void gives particles addresses, nothing more.

Aristotle famously claimed that "Nature abhors a vacuum," but I'm pretty sure a more correct translation would be "Nature abhors a void." Isaac Newton appeared to agree when he wrote in a letter to Richard Bentley:

> ... that one Body may act upon another at a Distance thro' a *Vacuum*, without the Mediation of any thing else, by and through which their Action and Force may be conveyed from one to another, is to me so great an Absurdity, that I believe no Man who has in philosophical Matters a competent Faculty of thinking, can ever fall into it.

But in Newton's masterpiece, the *Principia*, the players are bodies that exert forces on one another. Space, the stage, is an empty receptacle. It has no life of its own. In Newtonian physics, vacuum is a void.

That Newtonian framework worked brilliantly for nearly two centuries, as Newton's equations for gravity went from triumph to triumph, and (at first) the analogous ones for electric and magnetic forces seemed to do so as well. But in the 19th century, as people investigated the phenomena of electricity and magnetism more closely, Newton-style equations proved inadequate. In James Clerk Maxwell's equations, the fruit of that work, electromagnetic fields—not separated bodies—are the primary objects of reality.

Quantum theory amplified Maxwell's revolution. According to quantum theory, particles are merely bubbles of froth, kicked up by underlying fields. Photons, for example, are disturbances in electromagnetic fields.

As a young scientist, Feynman found that view too artificial. He wanted to bring back Newton's approach and work directly with the particles we actually perceive. In doing so, he hoped to challenge hidden assumptions and reach a simpler description of nature—and to avoid a big problem that the switch to quantum fields had created.

II.

In quantum theory, fields have a lot of spontaneous activity. They fluctuate in intensity and direction. And while the average value of the electric field in a vacuum is zero, the average value of its square is not zero. That's significant because the energy density in an electric field is proportional to the field's square. The energy density value, in fact, is infinite.

The spontaneous activity of quantum fields goes by several different names: quantum fluctuations, virtual particles or zero-point motion. There are subtle differences in the connotations of these expressions, but they all refer to the same phenomenon. Whatever you call it, the activity involves energy. Lots of energy—in fact, an infinite amount.

For most purposes we can leave that disturbing infinity out of consideration. Only changes in energy are observable. And because zero-point motion is an intrinsic characteristic of quantum fields, *changes* in energy, in response to external events, are generally finite. We can calculate them. They give rise to some very interesting effects, such as the Lamb shift of atomic spectral lines and the Casimir force between neutral conducting plates, which have been observed experimentally.[2] Far from being problematic, those effects are triumphs for quantum field theory.

The exception is gravity. Gravity responds to all kinds of energy, whatever form that energy may take. So the infinite energy density associated with the activity of quantum fields, present even in a vacuum, becomes a big problem when we consider its effect on gravity.

In principle, those quantum fields should make the vacuum heavy. Yet experiments tell us that the gravitational pull of the vacuum is quite small. Until recently—see more on this below—we thought it was zero.

Perhaps Feynman's conceptual switch from fields to particles would avoid the problem.

III.

Feynman started from scratch, drawing pictures whose stick-figure lines show links of influence between particles. The first published Feynman diagram, shown in figure 1.1, appeared in *Physical Review* in 1949.[3]

To understand how one electron influences another, using Feynman diagrams, you have to imagine that the electrons, as they move through space and evolve in time, exchange a photon, here labeled "virtual quantum." This is the simplest possibility. It is also possible to exchange two or more photons, and Feynman made similar diagrams for that. Those diagrams contribute another piece to the answer, modifying the classical Coulomb force law. By sprouting another squiggle, and letting it extend freely into the future, you represent how an electron radiates a photon. And so, step by step, you can describe complex physical processes, assembled like Tinkertoys from very simple ingredients.

Feynman diagrams look to be pictures of processes that happen in space and time, and in a sense they are, but they should not be interpreted too

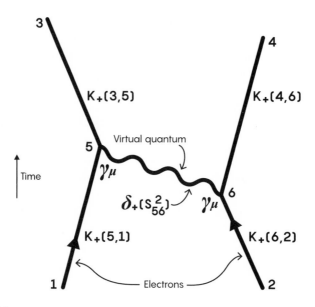

FIGURE 1.1
Two electrons exchange a photon.

literally. What they show are not rigid geometric trajectories, but more flexible, "topological" constructions, reflecting quantum uncertainty. In other words, you can be quite sloppy about the shape and configuration of the lines and squiggles, as long as you get the connections right.

Feynman found that he could attach a simple mathematical formula to each diagram. The formula expresses the likelihood of the process the diagram depicts. He found that in simple cases he got the same answers that people had obtained much more laboriously using fields when they let froth interact with froth.

That's what Feynman meant when he said, "There's nothing there." By removing the fields, he'd gotten rid of their contribution to gravity, which had led to absurdities. He thought he'd found a new approach to fundamental interactions that was not only simpler than the conventional one, but also sounder. It was a beautiful new way to think about fundamental processes.

IV.

Sadly, first appearances proved deceptive. As he worked things out further, Feynman discovered that his approach had a similar problem to the one it was supposed to solve. You can see this in the pictures below. We can draw

Feynman diagrams that are completely self-contained, without particles to initiate the events (or to flow out from them). These so-called disconnected graphs, or vacuum bubbles, are the Feynman diagram analogue of zero-point motion. You can draw diagrams for how virtual quanta affect gravitons, and thereby rediscover the morbid obesity of "empty" space.

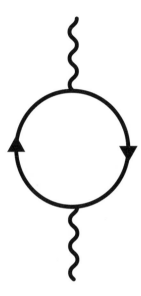

FIGURE 1.2
A graviton encounters a quantum fluctuation.

More generally, as he worked things out further, Feynman gradually realized—and then proved—that his diagram method is not a true alternative to the field approach, but rather an approximation to it. To Feynman, that came as a bitter disappointment.

Yet Feynman diagrams remain a treasured asset in physics, because they often provide good approximations to reality. Plus, they're easy (and fun) to work with. They help us bring our powers of visual imagination to bear on worlds we can't actually see.

The calculations that eventually got me a Nobel Prize in 2004 would have been literally unthinkable without Feynman diagrams, as would my calculations that established a route to production and observation of the Higgs particle.

On that day in Santa Barbara, citing those examples, I told Feynman how important his diagrams had been to me in my work. He seemed pleased,

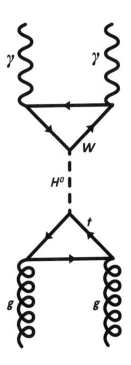

FIGURE 1.3
One way that the Higgs particle can be produced and then decay into daughter particles.

though he could hardly have been surprised at his diagrams' importance. "Yeah, that's the good part, seeing people use them, seeing them everywhere," he replied with a wink.

V.

The Feynman diagram representation of a process is most useful when a few relatively simple diagrams supply most of the answer. That is the regime physicists call "weak coupling," where each additional complicating line is relatively rare. That is almost always the case for photons in quantum electrodynamics (QED), the application Feynman originally had in mind. QED covers most of atomic physics, chemistry and materials science, so it's an amazing achievement to capture its essence in a few squiggles.

As an approach to the strong nuclear force, however, this strategy fails. Here the governing theory is quantum chromodynamics (QCD). The QCD analogues of photons are particles called color gluons, and their coupling is not weak. Usually, when we do a calculation in QCD, a host of complicated Feynman diagrams—festooned with many gluon lines—make important contributions to the answer. It's impractical (and probably impossible) to add them all up.

On the other hand, with modern computers we can go back to the truly fundamental field equations and calculate fluctuations in the quark and gluon fields directly. This approach gives beautiful pictures of another kind.

In recent years this direct approach, carried out on banks of supercomputers, has led to successful calculations of the masses of protons and neutrons. In the coming years it will revolutionize our quantitative understanding of nuclear physics over a broad front.

VI.

The puzzle Feynman thought he'd solved is still with us, though it has evolved in many ways.

The biggest change is that people have now measured the density of vacuum more precisely, and discovered that it does *not* vanish. It is the so-called "dark energy." (Dark energy is essentially—up to a numerical factor—the same thing Einstein called the "cosmological constant.") If you average it over the entire universe, you find that dark energy contributes about 70 percent of the total mass in the universe.

That sounds impressive, but for physicists the big puzzle that remains is why its density is as *small* as it is. For one thing, you'll remember, it was supposed to be infinite, due to the contribution of fluctuating fields. One bit of possible progress is that now we know a way to escape that infinity. It turns out that for one class of fields—technically, the fields associated with particles called bosons—the energy density is positive infinity, while for another class of fields—those associated with particles called fermions—the energy density is negative infinity. So if the universe contains an artfully balanced mix of bosons and fermions, the infinities can cancel. Supersymmetric theories, which also have several other attractive features, achieve that cancellation.

Another thing we've learned is that in addition to fluctuating fields, the vacuum contains nonfluctuating fields, often called "condensates." One such condensate is the so-called sigma condensate; another is the Higgs condensate. Those two are firmly established; there may be many others yet to be discovered. If you want to think of a familiar analogue, imagine Earth's magnetic or gravitational field, elevated to cosmic proportions (and freed of Earth). These condensates should also weigh something. Indeed, simple estimates of their density give values far larger than that of the observed dark energy.

We're left with an estimate of the dark energy that is finite (maybe), but poorly determined theoretically and, on the face of it, much too big. Presumably there are additional cancellations we don't know about. The most popular idea, at present, is that the smallness of the dark energy is a kind of rare accident, which happens to occur in our particular corner of the multiverse. Though unlikely a priori, it is necessary for our existence, and therefore what we are fated to observe.

That story, I'm afraid, is not nearly so elegant as Feynman's "There's nothing there!" Let's hope we can find a better one.

II WHAT IS QUANTUM REALITY, REALLY?

A JEWEL AT THE HEART OF QUANTUM PHYSICS

Natalie Wolchover

P hysicists have discovered a jewel-like geometric object that dramatically simplifies calculations of particle interactions and challenges the notion that space and time are fundamental components of reality.

"This is completely new and very much simpler than anything that has been done before," said Andrew Hodges, a mathematical physicist at Oxford University who has been following the work.

The revelation that particle interactions, the most basic events in nature, may be consequences of geometry significantly advances a decades-long effort to reformulate quantum field theory, the body of laws describing elementary particles and their interactions. Interactions that were previously calculated with mathematical formulas thousands of terms long can now be described by computing the volume of the corresponding jewel-like "amplituhedron," which yields an equivalent one-term expression.

"The degree of efficiency is mind-boggling," said Jacob Bourjaily, a theoretical physicist at the Niels Bohr Institute in Copenhagen and one of the researchers who developed the new idea. "You can easily do, on paper, computations that were infeasible even with a computer before."

The new geometric version of quantum field theory could also facilitate the search for a theory of quantum gravity that would seamlessly connect the large- and small-scale pictures of the universe. Attempts thus far to incorporate gravity into the laws of physics at the quantum scale have run up against nonsensical infinities and deep paradoxes. The amplituhedron, or a similar geometric object, could help by removing two deeply rooted principles of physics: locality and unitarity.

"Both are hard-wired in the usual way we think about things," said Nima Arkani-Hamed of the Institute for Advanced Study, who is the lead author of the new work.[1] "Both are suspect."

Locality is the notion that particles can interact only from adjoining positions in space and time. And unitarity holds that the probabilities of all possible outcomes of a quantum mechanical interaction must add up to one. The concepts are the central pillars of quantum field theory in its original form, but in certain situations involving gravity, both break down, suggesting neither is a fundamental aspect of nature.

In keeping with this idea, the new geometric approach to particle interactions removes locality and unitarity from its starting assumptions. The amplituhedron is not built out of space-time and probabilities; these properties merely arise as consequences of the jewel's geometry. The usual picture of space and time, and particles moving around in them, is a construct.

"It's a better formulation that makes you think about everything in a completely different way," said David Skinner, a theoretical physicist at Cambridge University.

The amplituhedron itself does not describe gravity. But Arkani-Hamed and his collaborators think there might be a related geometric object that does. Its properties would make it clear why particles appear to exist, and why they appear to move in three dimensions of space and to change over time.

Because "we know that ultimately, we need to find a theory that doesn't have" unitarity and locality, Bourjaily said, "it's a starting point to ultimately describing a quantum theory of gravity."

CLUNKY MACHINERY

The amplituhedron looks like an intricate, multifaceted jewel in higher dimensions. Encoded in its volume are the most basic features of reality that can be calculated, "scattering amplitudes," which represent the likelihood that a certain set of particles will turn into certain other particles upon colliding. These numbers are what particle physicists calculate and test to high precision at particle accelerators like the Large Hadron Collider in Switzerland.

The 70-year-old method for calculating scattering amplitudes—a major innovation at the time—was pioneered by the Nobel Prize-winning physicist Richard Feynman. He sketched line drawings of all the ways a scattering process could occur and then summed the likelihoods of the different drawings. The simplest Feynman diagrams look like trees: The particles involved in a collision come together like roots, and the particles that result shoot out like branches. More complicated diagrams have loops, where colliding particles turn into unobservable "virtual particles" that interact with each other before branching out as real final products. There are diagrams with one loop, two loops, three loops and so on—increasingly baroque iterations of the scattering process that contribute progressively less to its total amplitude. Virtual particles are never observed in nature, but they were considered mathematically necessary for unitarity—the requirement that probabilities sum to one.

"The number of Feynman diagrams is so explosively large that even computations of really simple processes weren't done until the age of computers," Bourjaily said. A seemingly simple event, such as two subatomic particles called gluons colliding to produce four less energetic gluons (which happens billions of times a second during collisions at the Large Hadron Collider), involves 220 diagrams, which collectively contribute thousands of terms to the calculation of the scattering amplitude.

In 1986, it became apparent that Feynman's apparatus was a Rube Goldberg machine.

To prepare for the construction of the Superconducting Super Collider in Texas (a project that was later canceled), theorists wanted to calculate the scattering amplitudes of known particle interactions to establish a background against which interesting or exotic signals would stand out. But even 2-gluon to 4-gluon processes were so complex, a group of physicists had written two years earlier, "that they may not be evaluated in the foreseeable future."

Stephen Parke and Tomasz Taylor, theorists at Fermi National Accelerator Laboratory in Illinois, took that statement as a challenge. Using a few mathematical tricks, they managed to simplify the 2-gluon to 4-gluon amplitude calculation from several billion terms to a 9-page-long formula, which a 1980s supercomputer could handle. Then, based on a pattern they observed in the scattering amplitudes of other gluon interactions, Parke and Taylor guessed a simple one-term expression for the amplitude. It was, the computer verified, equivalent to the 9-page formula. In other words, the traditional machinery of quantum field theory, involving hundreds of Feynman diagrams worth thousands of mathematical terms, was obfuscating something much simpler. As Bourjaily put it: "Why are you summing up millions of things when the answer is just one function?"

"We knew at the time that we had an important result," Parke said. "We knew it instantly. But what to do with it?"

THE AMPLITUHEDRON

The message of Parke and Taylor's single-term result took decades to interpret. "That one-term, beautiful little function was like a beacon for the next 30 years," Bourjaily said. It "really started this revolution."

In the mid-2000s, more patterns emerged in the scattering amplitudes of particle interactions, repeatedly hinting at an underlying, coherent mathematical structure behind quantum field theory. Most important was a set of formulas called the BCFW recursion relations, named for Ruth Britto, Freddy Cachazo, Bo Feng and Edward Witten. Instead of describing scattering processes in terms of familiar variables like position and time and depicting them in thousands of Feynman diagrams, the BCFW relations are best couched in terms of strange variables called "twistors," and particle interactions can be captured in a handful of associated twistor diagrams. The relations gained rapid adoption as tools for computing scattering amplitudes relevant to experiments, such as collisions at the Large Hadron Collider. But their simplicity was mysterious.

"The terms in these BCFW relations were coming from a different world, and we wanted to understand what that world was," Arkani-Hamed said in 2013. "That's what drew me into the subject five years ago."

With the help of leading mathematicians such as Pierre Deligne, Arkani-Hamed and his collaborators discovered that the recursion relations and associated twistor diagrams corresponded to a well-known geometric object. In fact, as detailed in a paper posted to arXiv.org in December 2012 by Arkani-Hamed, Bourjaily, Cachazo, Alexander Goncharov, Alexander Post-nikov and Jaroslav Trnka, the twistor diagrams gave instructions for calculating the volume of pieces of this object, called the positive Grassmannian.[2]

Named for Hermann Grassmann, a 19th-century German linguist and mathematician who studied its properties, "the positive Grassmannian is the slightly more grown-up cousin of the inside of a triangle," Arkani-Hamed explained. Just as the inside of a triangle is a region in a two-dimensional space bounded by intersecting lines, the simplest case of the positive Grass-mannian is a region in an N-dimensional space bounded by intersecting planes. (N is the number of particles involved in a scattering process.)

It was a geometric representation of real particle data, such as the likelihood that two colliding gluons will turn into four gluons. But something was still missing.

The physicists hoped that the amplitude of a scattering process would emerge purely and inevitably from geometry, but locality and unitarity were dictating which pieces of the positive Grassmannian to add together to get it. They wondered whether the amplitude was "the answer to some particular mathematical question," said Trnka, a theoretical physicist at the University of California, Davis. "And it is," he said.

Arkani-Hamed and Trnka discovered that the scattering amplitude equals the volume of a brand-new mathematical object—the amplituhedron. The details of a particular scattering process dictate the dimensionality and facets of the corresponding amplituhedron. The pieces of the positive

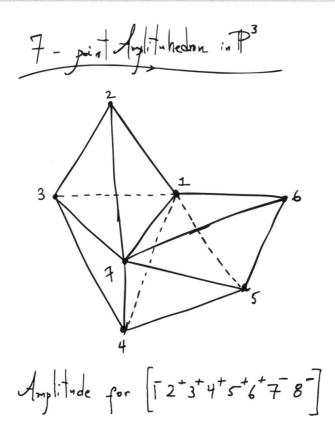

FIGURE 2.1

A sketch of the amplituhedron representing an 8-gluon particle interaction. Using Feynman diagrams, the same calculation would take roughly 500 pages of algebra. Courtesy of Nima Arkani-Hamed.

Grassmannian that were being calculated with twistor diagrams and then added together by hand were building blocks that fit together inside this jewel, just as triangles fit together to form a polygon.

Like the twistor diagrams, the Feynman diagrams are another way of computing the volume of the amplituhedron piece by piece, but they are much less efficient. "They are local and unitary in space-time, but they are not necessarily very convenient or well-adapted to the shape of this jewel itself," Skinner said. "Using Feynman diagrams is like taking a Ming vase and smashing it on the floor."

Arkani-Hamed and Trnka have been able to calculate the volume of the amplituhedron directly in some cases, without using twistor diagrams to compute the volumes of its pieces. They have also found a "master amplituhedron" with an infinite number of facets, analogous to a circle in 2-D, which has an infinite number of sides. Its volume represents, in theory, the total amplitude of all physical processes. Lower-dimensional amplituhedra, which correspond to interactions between finite numbers of particles, live on the faces of this master structure.

"They are very powerful calculational techniques, but they are also incredibly suggestive," Skinner said. "They suggest that thinking in terms of space-time was not the right way of going about this."

QUEST FOR QUANTUM GRAVITY

The seemingly irreconcilable conflict between gravity and quantum field theory enters crisis mode in black holes. Black holes pack a huge amount of mass into an extremely small space, making gravity a major player at the quantum scale, where it can usually be ignored. Inevitably, either locality or unitarity is the source of the conflict.

"We have indications that both ideas have got to go," Arkani-Hamed said. "They can't be fundamental features of the next description," such as a theory of quantum gravity.

String theory, a framework that treats particles as invisibly small, vibrating strings, is one candidate for a theory of quantum gravity that seems to hold up in black hole situations, but its relationship to reality is unproven—or at least confusing. Recently, a strange duality has been found between string theory and quantum field theory, indicating that the former (which includes gravity) is mathematically equivalent to the latter (which does not) when the two theories describe the same event as if it is taking place in different numbers of dimensions. No one knows quite what to make of this discovery. But the new amplituhedron research suggests space-time, and therefore dimensions, may be illusory anyway.

"We can't rely on the usual familiar quantum mechanical space-time pictures of describing physics," Arkani-Hamed said. "We have to learn new ways of talking about it. This work is a baby step in that direction."

Even without unitarity and locality, the amplituhedron formulation of quantum field theory does not yet incorporate gravity. But researchers are working on it. They say scattering processes that include gravity particles may be possible to describe with the amplituhedron, or with a similar geometric object. "It might be closely related but slightly different and harder to find," Skinner said.

Physicists must also prove that the new geometric formulation applies to the exact particles that are known to exist in the universe, rather than to the idealized quantum field theory they used to develop it, called maximally supersymmetric Yang-Mills theory. This model, which includes a "super-partner" particle for every known particle and treats space-time as flat, "just happens to be the simplest test case for these new tools," Bourjaily said. "The way to generalize these new tools to [other] theories is understood."

Beyond making calculations easier or possibly leading the way to quantum gravity, the discovery of the amplituhedron could cause an even more profound shift, Arkani-Hamed said. That is, giving up space and time as fundamental constituents of nature and figuring out how the Big Bang and cosmological evolution of the universe arose out of pure geometry.

"In a sense, we would see that change arises from the structure of the object," he said. "But it's not from the object changing. The object is basically timeless."

While more work is needed, many theoretical physicists are paying close attention to the new ideas.

The work is "very unexpected from several points of view," said Witten, a theoretical physicist at the Institute for Advanced Study. "The field is still developing very fast, and it is difficult to guess what will happen or what the lessons will turn out to be."

NEW SUPPORT FOR ALTERNATIVE QUANTUM VIEW

Dan Falk

O f the many counterintuitive features of quantum mechanics, perhaps the most challenging to our notions of common sense is that particles do not have locations until they are observed. This is exactly what the standard view of quantum mechanics, often called the Copenhagen interpretation, asks us to believe. Instead of the clear-cut positions and movements of Newtonian physics, we have a cloud of probabilities described by a mathematical structure known as a wave function. The wave function, meanwhile, evolves over time, its evolution governed by precise rules codified in something called the Schrödinger equation. The mathematics are clear enough; the actual whereabouts of particles, less so. Until a particle is observed, an act that causes the wave function to "collapse," we can say nothing about its location. Albert Einstein, among others, objected to this idea. As his biographer Abraham Pais wrote: "We often discussed his notions on objective reality. I recall that during one walk Einstein suddenly stopped, turned to me and asked whether I really believed that the moon exists only when I look at it."

But there's another view—one that's been around for almost a century—in which particles really do have precise positions at all times. This alternative view, known as pilot-wave theory or Bohmian mechanics, never became as popular as the Copenhagen view, in part because Bohmian mechanics implies that the world must be strange in other ways. In particular, a 1992 study claimed to crystalize certain bizarre consequences of Bohmian mechanics and in doing so deal it a fatal conceptual blow. The authors of that paper concluded that a particle following the laws of Bohmian mechanics would end up taking a trajectory that was so unphysical—even by the warped standards of quantum theory—that they described it as "surreal."[1]

Nearly a quarter-century later, a group of scientists has carried out an experiment in a Toronto laboratory that aims to test this idea. And if their results, first reported in 2016, hold up to scrutiny, the Bohmian view of quantum mechanics—less fuzzy but in some ways more strange than the traditional view—may be poised for a comeback.[2]

SAVING PARTICLE POSITIONS

Bohmian mechanics was worked out by Louis de Broglie in 1927 and again, independently, by David Bohm in 1952, who developed it further until his death in 1992. (It's also sometimes called the de Broglie–Bohm theory.) As with the Copenhagen view, there's a wave function governed by the Schrödinger equation. In addition, every particle has an actual, definite location, even when it's not being observed. Changes in the positions of the particles are given by another equation, known as the "pilot wave" equation (or "guiding equation"). The theory is fully deterministic; if you know the initial state of a system, and you've got the wave function, you can calculate where each particle will end up.

That may sound like a throwback to classical mechanics, but there's a crucial difference. Classical mechanics is purely "local"—stuff can affect other stuff only if it is adjacent to it (or via the influence of some kind of field, like an electric field, which can send impulses no faster than the speed of light). Quantum mechanics, in contrast, is inherently nonlocal. The best-known example of a nonlocal effect—one that Einstein himself considered, back in the 1930s—is when a pair of particles are connected in such a way that a measurement of one particle appears to affect the state of another, distant particle. The idea was ridiculed by Einstein as "spooky action at a distance." But hundreds of experiments, beginning in the 1980s, have confirmed that this spooky action is a very real characteristic of our universe.

In the Bohmian view, nonlocality is even more conspicuous. The trajectory of any one particle depends on what all the other particles described by the same wave function are doing. And, critically, the wave function has no geographic limits; it might, in principle, span the entire universe. Which means that the universe is weirdly interdependent, even across vast stretches of space. The wave function "combines—or binds—distant particles into a single irreducible reality," as Sheldon Goldstein, a mathematician and physicist at Rutgers University, has written.

The differences between Bohm and Copenhagen become clear when we look at the classic "double slit" experiment, in which particles (let's say electrons) pass through a pair of narrow slits, eventually reaching a screen where each particle can be recorded. When the experiment is carried out, the electrons behave like waves, creating on the screen a particular pattern called an "interference pattern." Remarkably, this pattern gradually emerges even if the electrons are sent one at a time, suggesting that each electron passes through both slits simultaneously.

Those who embrace the Copenhagen view have come to live with this state of affairs—after all, it's meaningless to speak of a particle's position until we measure it. Some physicists are drawn instead to the "many worlds" interpretation of quantum mechanics, in which observers in some universes see the electron go through the left slit, while those in other universes see it go through the right slit—which is fine, if you're comfortable with an infinite array of unseen universes.

By comparison, the Bohmian view sounds rather tame: The electrons act like actual particles, their velocities at any moment fully determined by the pilot wave, which in turn depends on the wave function. In this view, each electron is like a surfer: It occupies a particular place at every specific moment in time, yet its motion is dictated by the motion of a spread-out wave. Although each electron takes a fully determined path through just one slit, the pilot wave passes through both slits. The end result exactly matches the pattern one sees in standard quantum mechanics.

For some theorists, the Bohmian interpretation holds an irresistible appeal. "All you have to do to make sense of quantum mechanics is to say to yourself: When we talk about particles, we really mean particles. Then all the problems go away," said Goldstein. "Things have positions. They *are* somewhere. If you take that idea seriously, you're led almost immediately to Bohm. It's a far simpler version of quantum mechanics than what you find in the textbooks." Howard Wiseman, a physicist at Griffith University in Brisbane, Australia, said that the Bohmian view "gives you a pretty straightforward account of how the world is. ... You don't have to tie yourself into any sort of philosophical knots to say how things really are."

But not everyone feels that way, and over the years the Bohm view has struggled to gain acceptance, trailing behind Copenhagen and, these days, behind "many worlds" as well. A significant blow came with the paper known as "ESSW," an acronym built from the names of its four authors.[3] The ESSW paper claimed that particles can't follow simple Bohmian trajectories as they traverse the double-slit experiment. Suppose that someone placed a detector next to each slit, argued ESSW, recording which particle passed through which slit. ESSW showed that a photon could pass through the left slit and yet, in the Bohmian view, still end up being recorded as having passed through the right slit. This seemed impossible; the photons were deemed to follow "surreal" trajectories, as the ESSW paper put it.

The ESSW argument "was a striking philosophical objection" to the Bohmian view, said Aephraim Steinberg, a physicist at the University of Toronto. "It damaged my love for Bohmian mechanics."

But Steinberg has found a way to rekindle that love. In a paper published in *Science Advances*, Steinberg and his colleagues—the team includes Wiseman, in Australia, as well as five other Canadian researchers—describe what happened when they actually performed the ESSW experiment.[4] They found that the photon trajectories aren't surrealistic after all—or, more precisely, that the paths may seem surrealistic, but only if one fails to take into account the nonlocality inherent in Bohm's theory.

The experiment that Steinberg and his team conducted was analogous to the standard two-slit experiment. They used photons rather than electrons, and instead of sending those photons through a pair of slits, they passed through a beam splitter, a device that directs a photon along one of two paths, depending on the photon's polarization. The photons eventually reach a single-photon camera (equivalent to the screen in the traditional experiment) that records their final position. The question "Which of two slits did the particle pass through?" becomes "Which of two paths did the photon take?"

Importantly, the researchers used pairs of entangled photons rather than individual photons. As a result, they could interrogate one photon to gain information about the other. When the first photon passes through the beam splitter, the second photon "knows" which path the first one took. The team could then use information from the second photon to track the first photon's path. Each indirect measurement yields only an approximate value, but the scientists could average large numbers of measurements to reconstruct the trajectory of the first photon.

The team found that the photon paths do indeed appear to be surreal, just as ESSW predicted: A photon would sometimes strike one side of the screen, even though the polarization of the entangled partner said that the photon took the other route.

But can the information from the second photon be trusted? Crucially, Steinberg and his colleagues found that the answer to the question "Which path did the first photon take?" depends on when it is asked.

At first—in the moments immediately after the first photon passes through the beam splitter—the second photon is very strongly correlated with the first photon's path. "As one particle goes through the slit, the probe [the second photon] has a perfectly accurate memory of which slit it went through," Steinberg explained.

But the farther the first photon travels, the less reliable the second photon's report becomes. The reason is nonlocality. Because the two photons are entangled, the path that the first photon takes will affect the polarization of the second photon. By the time the first photon reaches the screen, the second photon's polarization is equally likely to be oriented one way as the other—thus giving it "no opinion," so to speak, as to whether the first photon took the first route or the second (the equivalent of knowing which of the two slits it went through).

The problem isn't that Bohm trajectories are surreal, said Steinberg. The problem is that the second photon says that Bohm trajectories are surreal—and, thanks to nonlocality, its report is not to be trusted. "There's no real contradiction in there," said Steinberg. "You just have to always bear in mind the nonlocality, or you miss something very important."

FASTER THAN LIGHT

Some physicists, unperturbed by ESSW, have embraced the Bohmian view all along and aren't particularly surprised by what Steinberg and his team found. There have been many attacks on the Bohmian view over the years, and "they all fizzled out because they had misunderstood what the Bohm approach was actually claiming," said Basil Hiley, a physicist at Birkbeck, University of London (formerly Birkbeck College), who collaborated with Bohm on his last book, *The Undivided Universe*. Owen Maroney, a physicist at the University of Oxford who was a student of Hiley's, described ESSW as "a terrible argument" that "did not present a novel challenge to de Broglie–Bohm." Not surprisingly, Maroney is excited by Steinberg's experimental results, which seem to support the view he's held all along. "It's a very interesting experiment," he said. "It gives a motivation for taking de Broglie–Bohm seriously."

On the other side of the Bohmian divide, Berthold-Georg Englert, one of the authors of ESSW (along with Marlan Scully, George Süssman and Herbert Walther), still describes their paper as a "fatal blow" to the Bohmian view. According to Englert, now at the National University of Singapore, the Bohm trajectories exist as mathematical objects but "lack physical meaning."

On a historical note, Einstein lived just long enough to hear about Bohm's revival of de Broglie's proposal—and he wasn't impressed, dismissing it as too simplistic to be correct. In a letter to physicist Max Born, in the spring of 1952, Einstein weighed in on Bohm's work:

> Have you noticed that Bohm believes (as de Broglie did, by the way, 25 years ago) that he is able to interpret the quantum theory in deterministic terms? That way seems too cheap to me. But you, of course, can judge this better than I.

But even for those who embrace the Bohmian view, with its clearly defined particles moving along precise paths, questions remain. Topping the list is an apparent tension with special relativity, which prohibits faster-than-light communication. Of course, as physicists have long noted, nonlocality of the sort associated with quantum entanglement does not allow for faster-than-light signaling (thus incurring no risk of the grandfather paradox or other violations of causality). Even so, many physicists feel that more clarification is needed, especially given the prominent role of nonlocality in the Bohmian view. The apparent dependence of what happens *here* on what may be happening *there* cries out for an explanation.

"The universe seems to like talking to itself faster than the speed of light," said Steinberg. "I could understand a universe where nothing can go faster than light, but a universe where the internal workings operate faster than light, and yet we're forbidden from ever making use of that at the macroscopic level—it's very hard to understand."

ENTANGLEMENT MADE SIMPLE

Frank Wilczek

An aura of glamorous mystery attaches to the concept of quantum entanglement, and also to the (somehow) related claim that quantum theory requires "many worlds." Yet in the end those are, or should be, scientific ideas, with down-to-earth meanings and concrete implications. Here I'd like to explain the concepts of entanglement and many worlds as simply and clearly as I know how.

I.

Entanglement is often regarded as a uniquely quantum-mechanical phenomenon, but it is not. In fact, it is enlightening, though somewhat unconventional, to consider a simple non quantum (or "classical") version of entanglement first. This enables us to pry the subtlety of entanglement itself apart from the general oddity of quantum theory.

Entanglement arises in situations where we have partial knowledge of the state of two systems. For example, our systems can be two objects that we'll call c-ons. The "c" is meant to suggest "classical," but if you'd prefer to have something specific and pleasant in mind, you can think of our c-ons as cakes.

Our c-ons come in two shapes, square or circular, which we identify as their possible states. Then the four possible joint states, for two c-ons, are (square, square), (square, circle), (circle, square), (circle, circle). The following tables show two examples of what the probabilities could be for finding the system in each of those four states.

We say that the c-ons are "independent" if knowledge of the state of one of them does not give useful information about the state of the other. Our first table has this property. If the first c-on (or cake) is square, we're still in the dark about the shape of the second. Similarly, the shape of the second does not reveal anything useful about the shape of the first.

Independent   ## Entangled

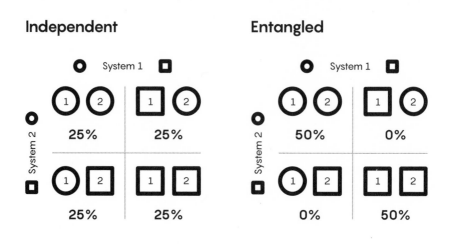

FIGURE 2.2

On the other hand, we say our two c-ons are entangled when information about one improves our knowledge of the other. Our second table
demonstrates extreme entanglement. In that case, whenever the first c-on
is circular, we know the second is circular too. And when the first c-on is
square, so is the second. Knowing the shape of one, we can infer the shape
of the other with certainty.

The quantum version of entanglement is essentially the sawme
phenomenon—that is, lack of independence. In quantum theory, states
are described by mathematical objects called wave functions. The rules connecting wave functions to physical probabilities introduce very interesting complications, as we will discuss, but the central concept of entangled
knowledge, which we have seen already for classical probabilities, carries
over.

Cakes don't count as quantum systems, of course, but entanglement
between quantum systems arises naturally—for example, in the aftermath
of particle collisions. In practice, unentangled (independent) states are rare
exceptions, for whenever systems interact, the interaction creates correlations between them.

Consider, for example, molecules. They are composites of subsystems,
namely electrons and nuclei. A molecule's lowest energy state, in which it
is most usually found, is a highly entangled state of its electrons and nuclei,
for the positions of those constituent particles are by no means independent. As the nuclei move, the electrons move with them.

Returning to our example: If we write $\Phi_\blacksquare$, $\Phi_\bullet$ for the wave functions describing system 1 in its square or circular states, and $\Psi_\blacksquare$, $\Psi_\bullet$ for the wave functions describing system 2 in its square or circular states, then in our working example the overall states will be

Independent: $\Phi_\blacksquare\,\Psi_\blacksquare + \Phi_\blacksquare\,\Psi_\bullet + \Phi_\bullet\,\Psi_\blacksquare + \Phi_\bullet\,\Psi_\bullet$

Entangled: $\Phi_\blacksquare\,\Psi_\blacksquare + \Phi_\bullet\,\Psi_\bullet$

We can also write the independent version as

$(\Phi_\blacksquare + \Phi_\bullet)(\Psi_\blacksquare + \Psi_\bullet)$

Note how in this formulation the parentheses clearly separate systems 1 and 2 into independent units.

There are many ways to create entangled states. One way is to make a measurement of your (composite) system that gives you partial information. We can learn, for example, that the two systems have conspired to have the same shape, without learning exactly what shape they have. This concept will become important later.

The more distinctive consequences of quantum entanglement, such as the Einstein-Podolsky-Rosen (EPR) and Greenberger-Horne-Zeilinger (GHZ) effects, arise through its interplay with another aspect of quantum theory called "complementarity." To pave the way for discussion of EPR and GHZ, let me now introduce complementarity.

Previously, we imagined that our c-ons could exhibit two shapes (square and circle). Now we imagine that it can also exhibit two colors—gray and black. If we were speaking of classical systems, like cakes, this added property would imply that our c-ons could be in any of four possible states: a gray square, a gray circle, a black square or a black circle.

Yet for a quantum cake—a quake, perhaps, or (with more dignity) a q-on—the situation is profoundly different. The fact that a q-on can exhibit, in different situations, different shapes or different colors does not necessarily mean that it possesses both a shape and a color simultaneously. In fact, that "common sense" inference, which Einstein insisted should be part of any acceptable notion of physical reality, is inconsistent with experimental facts, as we'll see shortly.

We can measure the shape of our q-on, but in doing so we lose all information about its color. Or we can measure the color of our q-on, but in doing so we lose all information about its shape. What we cannot do, according to quantum theory, is measure both its shape and its color simultaneously. No one view of physical reality captures all its aspects; one must take into account many different, mutually exclusive views, each offering valid but partial insight. This is the heart of complementarity, as Niels Bohr formulated it.

As a consequence, quantum theory forces us to be circumspect in assigning physical reality to individual properties. To avoid contradictions, we must admit that:

1. A property that is not measured need not exist.
2. Measurement is an active process that alters the system being measured.

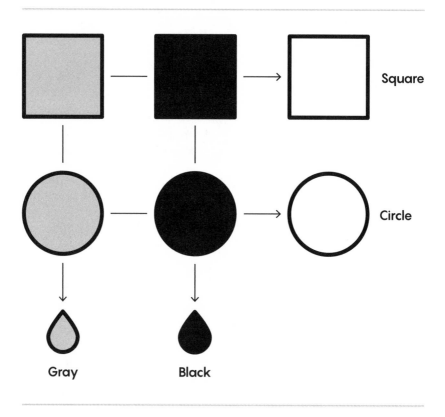

FIGURE 2.3

II.

Now I will describe two classic—though far from classical!—illustrations of quantum theory's strangeness. Both have been checked in rigorous experiments. (In the actual experiments, people measure properties like the angular momentum of electrons rather than shapes or colors of cakes.)

Albert Einstein, Boris Podolsky and Nathan Rosen (EPR) described a startling effect that can arise when two quantum systems are entangled. The

EPR effect marries a specific, experimentally realizable form of quantum entanglement with complementarity.

An EPR pair consists of two q-ons, each of which can be measured either for its shape or for its color (but not for both). We assume that we have access to many such pairs, all identical, and that we can choose which measurements to make of their components. If we measure the shape of one member of an EPR pair, we find it is equally likely to be square or circular. If we measure the color, we find it is equally likely to be gray or black.

The interesting effects, which EPR considered paradoxical, arise when we make measurements of both members of the pair. When we measure both members for color, or both members for shape, we find that the results always agree. Thus if we find that one is gray, and later measure the color of the other, we will discover that it too is gray, and so forth. On the other hand, if we measure the shape of one, and then the color of the other, there is no correlation. Thus if the first is square, the second is equally likely to be gray or to be black.

We will, according to quantum theory, get those results even if great distances separate the two systems, and the measurements are performed nearly simultaneously. The choice of measurement in one location appears to be affecting the state of the system in the other location. This "spooky action at a distance," as Einstein called it, might seem to require transmission of information—in this case, information about what measurement was performed—at a rate faster than the speed of light.

But does it? Until I *know* the result you obtained, I don't know what to expect. I gain useful information when I learn the result you've measured, not at the moment you measure it. And any message revealing the result you measured must be transmitted in some concrete physical way, slower (presumably) than the speed of light.

Upon deeper reflection, the paradox dissolves further. Indeed, let us consider again the state of the second system, given that the first has been measured to be gray. If we choose to measure the second q-on's color, we will surely get gray. But as we discussed earlier, when introducing complementarity, if we choose to measure a q-on's shape, when it is in the "red" state, we will have equal probability to find a square or a circle. Thus, far from introducing a paradox, the EPR outcome is logically forced. It is, in essence, simply a repackaging of complementarity.

Nor is it paradoxical to find that distant events are correlated. After all, if I put each member of a pair of gloves in boxes, and mail them to opposite sides of the earth, I should not be surprised that by looking inside one box I can determine the handedness of the glove in the other. Similarly, in all known cases the correlations between an EPR pair must be imprinted when its members are close together, though of course they can survive subsequent separation, as though they had memories. Again, the peculiarity of EPR is not correlation as such, but its possible embodiment in complementary forms.

III.

Daniel Greenberger, Michael Horne and Anton Zeilinger discovered another brilliantly illuminating example of quantum entanglement.[1] It involves three of our q-ons, prepared in a special, entangled state (the GHZ state). We distribute the three q-ons to three distant experimenters. Each experimenter chooses, independently and at random, whether to measure shape or color, and records the result. The experiment gets repeated many times, always with the three q-ons starting out in the GHZ state.

Each experimenter, separately, finds maximally random results. When she measures a q-on's shape, she is equally likely to find a square or a circle; when she measures its color, gray or black are equally likely. So far, so mundane.

But later, when the experimenters come together and compare their measurements, a bit of analysis reveals a stunning result. Let us call square shapes and gray colors "good," and circular shapes and black colors "evil." The experimenters discover that whenever two of them chose to measure shape but the third measured color, they found that exactly 0 or 2 results were "evil" (that is, circular or black). But when all three chose to measure color, they found that exactly 1 or 3 measurements were evil. That is what quantum mechanics predicts, and that is what is observed.

So: Is the quantity of evil even or odd? Both possibilities are realized, with certainty, in different sorts of measurements. We are forced to reject the question. It makes no sense to speak of the quantity of evil in our system, independent of how it is measured. Indeed, it leads to contradictions.

The GHZ effect is, in the physicist Sidney Coleman's words, "quantum mechanics in your face." It demolishes a deeply embedded prejudice, rooted in everyday experience, that physical systems have definite properties, independent of whether those properties are measured. For if they did, then the balance between good and evil would be unaffected by measurement choices. Once internalized, the message of the GHZ effect is unforgettable and mind-expanding.

IV.

Thus far we have considered how entanglement can make it impossible to assign unique, independent states to several q-ons. Similar considerations apply to the evolution of a single q-on in time.

We say we have "entangled histories" when it is impossible to assign a definite state to our system at each moment in time. Similarly to how we got conventional entanglement by eliminating some possibilities, we can create entangled histories by making measurements that gather partial information about what happened. In the simplest entangled histories, we have just one q-on, which we monitor at two different times. We can imagine situations where we determine that the shape of our q-on was either square at both times or that it was circular at both times, but that our observations leave both alternatives in play. This is a quantum temporal analogue of the simplest entanglement situations illustrated above.

Using a slightly more elaborate protocol we can add the wrinkle of complementarity to this system, and define situations that bring out the "many worlds" aspect of quantum theory. Thus our q-on might be prepared in the gray state at an earlier time, and measured to be in the black state at a subsequent time. As in the simple examples above, we cannot consistently assign our q-on the property of color at intermediate times; nor does it have a determinate shape. Histories of this sort realize, in a limited but controlled and precise way, the intuition that underlies the many worlds picture of quantum mechanics. A definite state can branch into mutually contradictory historical trajectories that later come together.

Erwin Schrödinger, a founder of quantum theory who was deeply skeptical of its correctness, emphasized that the evolution of quantum systems naturally leads to states that might be measured to have grossly different properties. His "Schrödinger cat" states, famously, scale up quantum uncertainty into questions about feline mortality. Prior to measurement, as we've seen in our examples, one cannot assign the property of life (or death) to the cat. Both—or neither—coexist within a netherworld of possibility.

Everyday language is ill suited to describe quantum complementarity, in part because everyday experience does not encounter it. Practical cats interact with surrounding air molecules, among other things, in very different ways depending on whether they are alive or dead, so in practice the measurement gets made automatically, and the cat gets on with its life (or death). But entangled histories describe q-ons that are, in a real sense, Schrödinger kittens. Their full description requires, at intermediate times, that we take both of two contradictory property-trajectories into account.

The controlled experimental realization of entangled histories is delicate because it requires we gather partial information about our q-on. Conventional quantum measurements generally gather complete information at one time—for example, they determine a definite shape, or a definite color—rather than partial information spanning several times. But it can be done—indeed, without great technical difficulty. In this way we can give definite mathematical and experimental meaning to the proliferation of "many worlds" in quantum theory, and demonstrate its substantiality.

QUANTUM THEORY REBUILT FROM SIMPLE PHYSICAL PRINCIPLES

Philip Ball

S cientists have been using quantum theory for almost a century now, but embarrassingly they still don't know what it means. An informal poll taken at a 2011 conference on Quantum Physics and the Nature of Reality showed that there's still no consensus on what quantum theory says about reality—the participants remained deeply divided about how the theory should be interpreted.[1]

Some physicists just shrug and say we have to live with the fact that quantum mechanics is weird. So particles can be in two places at once, or communicate instantaneously over vast distances? Get over it. After all, the theory works fine. If you want to calculate what experiments will reveal about subatomic particles, atoms, molecules and light, then quantum mechanics succeeds brilliantly.

But some researchers want to dig deeper. They want to know why quantum mechanics has the form it does, and they are engaged in an ambitious program to find out. It is called quantum reconstruction, and it amounts to trying to rebuild the theory from scratch based on a few simple principles.

If these efforts succeed, it's possible that all the apparent oddness and confusion of quantum mechanics will melt away, and we will finally grasp what the theory has been trying to tell us. "For me, the ultimate goal is to prove that quantum theory is the only theory where our imperfect experiences allow us to build an ideal picture of the world," said Giulio Chiribella, a theoretical physicist at the University of Oxford.

There's no guarantee of success—no assurance that quantum mechanics really does have something plain and simple at its heart, rather than the abstruse collection of mathematical concepts used today. But even if quantum reconstruction efforts don't pan out, they might point the way to an equally tantalizing goal: getting beyond quantum mechanics itself to a still deeper theory. "I think it might help us move towards a theory of quantum gravity," said Lucien Hardy, a theoretical physicist at the Perimeter Institute for Theoretical Physics in Waterloo, Canada.

THE FLIMSY FOUNDATIONS OF QUANTUM MECHANICS

The basic premise of the quantum reconstruction game is summed up by the joke about the driver who, lost in rural Ireland, asks a passer-by how to get to Dublin. "I wouldn't start from here," comes the reply.

Where, in quantum mechanics, is "here"? The theory arose out of attempts to understand how atoms and molecules interact with light and other radiation, phenomena that classical physics couldn't explain. Quantum theory was empirically motivated, and its rules were simply ones that seemed to fit what was observed. It uses mathematical formulas that, while tried and trusted, were essentially pulled out of a hat by the pioneers of the theory in the early 20th century.

Take Erwin Schrödinger's equation for calculating the probabilistic properties of quantum particles. The particle is described by a "wave function" that encodes all we can know about it. It's basically a wavelike mathematical expression, reflecting the well-known fact that quantum particles can sometimes seem to behave like waves. Want to know the probability that the particle will be observed in a particular place? Just calculate the square of the wave function (or, to be exact, a slightly more complicated mathematical term), and from that you can deduce how likely you are to detect the particle there. The probability of measuring some of its other observable properties can be found by, crudely speaking, applying a mathematical function called an operator to the wave function.

But this so-called rule for calculating probabilities was really just an intuitive guess by the German physicist Max Born. So was Schrödinger's equation itself. Neither was supported by rigorous derivation. Quantum mechanics seems largely built of arbitrary rules like this, some of them—such as the mathematical properties of operators that correspond to observable properties of the system—rather arcane. It's a complex framework, but it's also an ad hoc patchwork, lacking any obvious physical interpretation or justification.

Compare this with the ground rules, or axioms, of Einstein's theory of special relativity, which was as revolutionary in its way as quantum mechanics. (Einstein launched them both, rather miraculously, in 1905.) Before Einstein, there was an untidy collection of equations to describe how light behaves from the point of view of a moving observer. Einstein dispelled the mathematical fog with two simple and intuitive principles: that the speed of light is constant, and that the laws of physics are the same for two observers moving at constant speed relative to one another. Grant these basic principles, and the rest of the theory follows. Not only are the axioms simple, but we can see at once what they mean in physical terms.

What are the analogous statements for quantum mechanics? The eminent physicist John Wheeler once asserted that if we really understood the central point of quantum theory, we would be able to state it in one simple sentence that anyone could understand. If such a statement exists, some quantum reconstructionists suspect that we'll find it only by rebuilding quantum theory from scratch: by tearing up the work of Bohr, Heisenberg and Schrödinger and starting again.

QUANTUM ROULETTE

One of the first efforts at quantum reconstruction was made in 2001 by Hardy, then at the University of Oxford.[2] He ignored everything that we typically associate with quantum mechanics, such as quantum jumps, wave-particle duality and uncertainty. Instead, Hardy focused on probability: specifically, the probabilities that relate the possible states of a system with the chance of observing each state in a measurement. Hardy found that these bare bones were enough to get all that familiar quantum stuff back again.

Hardy assumed that any system can be described by some list of properties and their possible values. For example, in the case of a tossed coin, the salient values might be whether it comes up heads or tails. Then he considered the possibilities for measuring those values definitively in a single observation. You might think any distinct state of any system can always be reliably distinguished (at least in principle) by a measurement or observation. And that's true for objects in classical physics.

In quantum mechanics, however, a particle can exist not just in distinct states, like the heads and tails of a coin, but in a so-called superposition—roughly speaking, a combination of those states. In other words, a quantum bit, or qubit, can be not just in the binary state of 0 or 1, but in a superposition of the two.

But if you make a measurement of that qubit, you'll only ever get a result of 1 or 0. That is the mystery of quantum mechanics, often referred to as the collapse of the wave function: Measurements elicit only one of the possible outcomes. To put it another way, a quantum object commonly has more options for measurements encoded in the wave function than can be seen in practice.

Hardy's rules governing possible states and their relationship to measurement outcomes acknowledged this property of quantum bits. In essence the rules were (probabilistic) ones about how systems can carry information and how they can be combined and interconverted.

Hardy then showed that the simplest possible theory to describe such systems is quantum mechanics, with all its characteristic phenomena such as wavelike interference and entanglement, in which the properties of different objects become interdependent. "Hardy's 2001 paper was the 'Yes, we can!' moment of the reconstruction program," Chiribella said. "It told us that in some way or another we can get to a reconstruction of quantum theory."

More specifically, it implied that the core trait of quantum theory is that it is inherently probabilistic. "Quantum theory can be seen as a generalized probability theory, an abstract thing that can be studied detached from its application to physics," Chiribella said. This approach doesn't address any underlying physics at all, but just considers how outputs are related to inputs: what we can measure given how a state is prepared (a so-called operational perspective). "What the physical system is is not specified and plays no role in the results," Chiribella said. These generalized probability theories are "pure syntax," he added—they relate states and measurements, just as linguistic syntax relates categories of words, without regard to what the words mean. In other words, Chiribella explained, generalized probability theories "are the syntax of physical theories, once we strip them of the semantics."

The general idea for all approaches in quantum reconstruction, then, is to start by listing the probabilities that a user of the theory assigns to each of the possible outcomes of all the measurements the user can perform on a system. That list is the "state of the system." The only other ingredients are the ways in which states can be transformed into one another, and the probability of the outputs given certain inputs. This operational approach to reconstruction "doesn't assume space-time or causality or anything, only a distinction between these two types of data," said Alexei Grinbaum, a philosopher of physics at the CEA Saclay in France.

To distinguish quantum theory from a generalized probability theory, you need specific kinds of constraints on the probabilities and possible outcomes of measurement. But those constraints aren't unique. So lots of possible theories of probability look quantum-like. How then do you pick out the right one?

"We can look for probabilistic theories that are similar to quantum theory but differ in specific aspects," said Matthias Kleinmann, a theoretical physicist at the University of the Basque Country in Bilbao, Spain. If you can then find postulates that select quantum mechanics specifically, he explained, you can "drop or weaken some of them and work out mathematically what other theories appear as solutions." Such exploration of what lies beyond quantum mechanics is not just academic doodling, for it's possible—indeed, likely—that quantum mechanics is itself just an approximation of a deeper theory. That theory might emerge, as quantum theory did from classical physics, from violations in quantum theory that appear if we push it hard enough.

BITS AND PIECES

Some researchers suspect that ultimately the axioms of a quantum reconstruction will be about information: what can and can't be done with it.[3] One such derivation of quantum theory based on axioms about information was proposed in 2010 by Chiribella, then working at the Perimeter Institute, and his collaborators Giacomo Mauro D'Ariano and Paolo Perinotti of the University of Pavia in Italy.[4] "Loosely speaking," explained Jacques Pienaar, a theoretical physicist at the International Institute of Physics in Natal, Brazil, "their principles state that information should be localized in space and time, that systems should be able to encode information about each other, and that every process should in principle be reversible, so that information is conserved." (In irreversible processes, by contrast, information is typically lost—just as it is when you erase a file on your hard drive.)

What's more, said Pienaar, these axioms can all be explained using ordinary language. "They all pertain directly to the elements of human experience, namely, what real experimenters ought to be able to do with the systems in their laboratories," he said. "And they all seem quite reasonable, so that it is easy to accept their truth." Chiribella and his colleagues showed that a system governed by these rules shows all the familiar quantum behaviors, such as superposition and entanglement.

One challenge is to decide what should be designated an axiom and what physicists should try to derive from the axioms. Take the quantum no-cloning rule, which is another of the principles that naturally arises from Chiribella's reconstruction. One of the deep findings of modern quantum theory, this principle states that it is impossible to make a duplicate of an arbitrary, unknown quantum state.

It sounds like a technicality (albeit a highly inconvenient one for scientists and mathematicians seeking to design quantum computers). But in an effort in 2002 to derive quantum mechanics from rules about what is permitted with quantum information, Jeffrey Bub of the University of Maryland and his colleagues Rob Clifton of the University of Pittsburgh and Hans Halvorson of Princeton University made no-cloning one of three fundamental axioms.[5] One of the others was a straightforward consequence of special relativity: You can't transmit information between two objects more quickly than the speed of light by making a measurement on one of the objects. The third axiom was harder to state, but it also crops up as a constraint on quantum information technology. In essence, it limits how securely a bit of information can be exchanged without being tampered with: The rule is a prohibition on what is called "unconditionally secure bit commitment."

These axioms seem to relate to the practicalities of managing quantum information. But if we consider them instead to be fundamental, and if we additionally assume that the algebra of quantum theory has a property called noncommutation, meaning that the order in which you do calculations matters (in contrast to the multiplication of two numbers, which can be done in any order), Clifton, Bub and Halvorson have shown that these rules too give rise to superposition, entanglement, uncertainty, nonlocality and so on: the core phenomena of quantum theory.

Another information-focused reconstruction was suggested in 2009 by Borivoje Dakić and Časlav Brukner, physicists at the University of Vienna.[6] They proposed three "reasonable axioms" having to do with information capacity: that the most elementary component of all systems can carry no more than one bit of information, that the state of a composite system made up of subsystems is completely determined by measurements on its subsystems, and that you can convert any "pure" state to another and back again (like flipping a coin between heads and tails).

Dakić and Brukner showed that these assumptions lead inevitably to classical and quantum-style probability, and to no other kinds. What's more, if you modify axiom three to say that states get converted continuously— little by little, rather than in one big jump—you get only quantum theory, not classical. (Yes, it really is that way round, contrary to what the "quantum jump" idea would have you expect—you can interconvert states of quantum spins by rotating their orientation smoothly, but you can't gradually convert a classical heads to a tails.) "If we don't have continuity, then we don't have quantum theory," Grinbaum said.

A further approach in the spirit of quantum reconstruction is called quantum Bayesianism, or QBism. Devised by Carlton Caves, Christopher Fuchs and Rüdiger Schack in the early 2000s, it takes the provocative position that the mathematical machinery of quantum mechanics has nothing to do with the way the world really is; rather, it is just the appropriate framework that lets us develop expectations and beliefs about the outcomes of our interventions. It takes its cue from the Bayesian approach to classical probability developed in the 18th century, in which probabilities stem from personal beliefs rather than observed frequencies. In QBism, quantum probabilities calculated by the Born rule don't tell us what we'll measure, but only what we should rationally expect to measure.

In this view, the world isn't bound by rules—or at least, not by quantum rules. Indeed, there may be no fundamental laws governing the way particles interact; instead, laws emerge at the scale of our observations. This possibility was considered by John Wheeler, who dubbed the scenario Law Without Law. It would mean that "quantum theory is merely a tool to make comprehensible a lawless slicing-up of nature," said Adán Cabello, a physicist at the University of Seville. Can we derive quantum theory from these premises alone?

"At first sight, it seems impossible," Cabello admitted—the ingredients seem far too thin, not to mention arbitrary and alien to the usual assumptions of science. "But what if we manage to do it?" he asked. "Shouldn't this shock anyone who thinks of quantum theory as an expression of properties of nature?"

MAKING SPACE FOR GRAVITY

In Hardy's view, quantum reconstructions have been almost too successful, in one sense: Various sets of axioms all give rise to the basic structure of quantum mechanics. "We have these different sets of axioms, but when you look at them, you can see the connections between them," he said. "They all seem reasonably good and are in a formal sense equivalent because they all give you quantum theory." And that's not quite what he'd hoped for. "When I started on this, what I wanted to see was two or so obvious, compelling axioms that would give you quantum theory and which no one would argue with."

So how do we choose between the options available? "My suspicion now is that there is still a deeper level to go to in understanding quantum theory," Hardy said. And he hopes that this deeper level will point beyond quantum theory, to the elusive goal of a quantum theory of gravity. "That's the next step," he said. Several researchers working on reconstructions now hope that its axiomatic approach will help us see how to pose quantum theory in a way that forges a connection with the modern theory of gravitation—Einstein's general relativity.

Look at the Schrödinger equation and you will find no clues about how to take that step. But quantum reconstructions with an "informational" flavor speak about how information-carrying systems can affect one another, a framework of causation that hints at a link to the space-time picture of general relativity. Causation imposes chronological ordering: An effect can't precede its cause. But Hardy suspects that the axioms we need to build quantum theory will be ones that embrace a lack of definite causal structure—no unique time-ordering of events—which he says is what we should expect when quantum theory is combined with general relativity. "I'd like to see axioms that are as causally neutral as possible, because they'd be better candidates as axioms that come from quantum gravity," he said.

Hardy first suggested that quantum-gravitational systems might show indefinite causal structure in 2007.[7] And in fact only quantum mechanics can display that. While working on quantum reconstructions, Chiribella was inspired to propose an experiment to create causal superpositions of quantum systems, in which there is no definite series of cause-and-effect events.[8] This experiment has now been carried out by Philip Walther's lab at the University of Vienna—and it might incidentally point to a way of making quantum computing more efficient.

"I find this a striking illustration of the usefulness of the reconstruction approach," Chiribella said. "Capturing quantum theory with axioms is not just an intellectual exercise. We want the axioms to do something useful for us—to help us reason about quantum theory, invent new communication protocols and new algorithms for quantum computers, and to be a guide for the formulation of new physics."

But can quantum reconstructions also help us understand the "meaning" of quantum mechanics? Hardy doubts that these efforts can resolve arguments about interpretation—whether we need many worlds or just one, for example. After all, precisely because the reconstructionist program is inherently "operational," meaning that it focuses on the "user experience"—probabilities about what we measure—it may never speak about the "underlying reality" that creates those probabilities.

"When I went into this approach, I hoped it would help to resolve these interpretational problems," Hardy admitted. "But I would say it hasn't." Cabello agrees. "One can argue that previous reconstructions failed to make quantum theory less puzzling or to explain where quantum theory comes from," he said. "All of them seem to miss the mark for an ultimate understanding of the theory." But he remains optimistic: "I still think that the right approach will dissolve the problems and we will understand the theory."

Maybe, Hardy said, these challenges stem from the fact that the more fundamental description of reality is rooted in that still undiscovered theory of quantum gravity. "Perhaps when we finally get our hands on quantum gravity, the interpretation will suggest itself," he said. "Or it might be worse!"

Right now, quantum reconstruction has few adherents—which pleases Hardy, as it means that it's still a relatively tranquil field. But if it makes serious inroads into quantum gravity, that will surely change. In the 2011 poll, about a quarter of the respondents felt that quantum reconstructions will lead to a new, deeper theory. A one-in-four chance certainly seems worth a shot.

Grinbaum thinks that the task of building the whole of quantum theory from scratch with a handful of axioms may ultimately be unsuccessful. "I'm now very pessimistic about complete reconstructions," he said. But, he suggested, why not try to do it piece by piece instead—to just reconstruct particular aspects, such as nonlocality or causality? "Why would one try to reconstruct the entire edifice of quantum theory if we know that it's made of different bricks?" he asked. "Reconstruct the bricks first. Maybe remove some and look at what kind of new theory may emerge."

"I think quantum theory as we know it will not stand," Grinbaum said. "Which of its feet of clay will break first is what reconstructions are trying to explore." He thinks that, as this daunting task proceeds, some of the most vexing and vague issues in standard quantum theory—such as the process of measurement and the role of the observer—will disappear, and we'll see that the real challenges are elsewhere. "What is needed is new mathematics that will render these notions scientific," he said. Then, perhaps, we'll understand what we've been arguing about for so long.

WHAT IS TIME?

TIME'S ARROW TRACED TO QUANTUM SOURCE

Natalie Wolchover

C offee cools, buildings crumble, eggs break and stars fizzle out in a universe that seems destined to degrade into a state of uniform drabness known as thermal equilibrium. The astronomer-philosopher Sir Arthur Eddington in 1927 cited the gradual dispersal of energy as evidence of an irreversible "arrow of time."

But to the bafflement of generations of physicists, the arrow of time does not seem to follow from the underlying laws of physics, which work the same going forward in time as in reverse. By those laws, it seemed that if someone knew the paths of all the particles in the universe and flipped them around, energy would accumulate rather than disperse: Tepid coffee would spontaneously heat up, buildings would rise from their rubble and sunlight would slink back into the sun.

"In classical physics, we were struggling," said Sandu Popescu, a professor of physics at the University of Bristol in the United Kingdom. "If I knew more, could I reverse the event, put together all the molecules of the egg that broke? Why am I relevant?"

Surely, he said, time's arrow is not steered by human ignorance. And yet, since the birth of thermodynamics in the 1850s, the only known approach for calculating the spread of energy was to formulate statistical distributions of the unknown trajectories of particles, and show that, over time, the ignorance smeared things out.

Now, physicists are unmasking a more fundamental source for the arrow of time: Energy disperses and objects equilibrate, they say, because of the way elementary particles become intertwined when they interact—a strange effect called "quantum entanglement."

"Finally, we can understand why a cup of coffee equilibrates in a room," said Tony Short, a quantum physicist at Bristol. "Entanglement builds up between the state of the coffee cup and the state of the room."

Popescu, Short and their colleagues Noah Linden and Andreas Winter reported the discovery in the journal *Physical Review E* in 2009, arguing that objects reach equilibrium, or a state of uniform energy distribution, within an infinite amount of time by becoming quantum mechanically entangled with their surroundings.[1] Similar results by Peter Reimann of the University of Bielefeld in Germany appeared several months earlier in *Physical Review Letters*.[2] Short and a collaborator strengthened the argument in 2012 by showing that entanglement causes equilibration within a finite time.[3] And, in work that was posted on the scientific preprint site arXiv.org in February 2014, two separate groups have taken the next step, calculating that most physical systems equilibrate rapidly, on time scales proportional to their size.[4,5] "To show that it's relevant to our actual physical world, the processes have to be happening on reasonable time scales," Short said.

The tendency of coffee—and everything else—to reach equilibrium is "very intuitive," said Nicolas Brunner, a quantum physicist at the University of Geneva. "But when it comes to explaining why it happens, this is the first time it has been derived on firm grounds by considering a microscopic theory."

If the new line of research is correct, then the story of time's arrow begins with the quantum mechanical idea that, deep down, nature is inherently uncertain. An elementary particle lacks definite physical properties and is defined only by probabilities of being in various states. For example, at a particular moment, a particle might have a 50 percent chance of spinning clockwise and a 50 percent chance of spinning counterclockwise. An experimentally tested theorem by the Northern Irish physicist John Bell says there is no "true" state of the particle; the probabilities are the only reality that can be ascribed to it.

Quantum uncertainty then gives rise to entanglement, the putative source of the arrow of time.

When two particles interact, they can no longer even be described by their own, independently evolving probabilities, called "pure states." Instead, they become entangled components of a more complicated probability distribution that describes both particles together. It might dictate, for example, that the particles spin in opposite directions. The system as a whole is in a pure state, but the state of each individual particle is "mixed" with that of its acquaintance. The two could travel light-years apart, and the spin of each would remain correlated with that of the other, a feature Albert Einstein famously described as "spooky action at a distance."

"Entanglement is in some sense the essence of quantum mechanics," or the laws governing interactions on the subatomic scale, Brunner said. The phenomenon underlies quantum computing, quantum cryptography and quantum teleportation.

The idea that entanglement might explain the arrow of time first occurred to Seth Lloyd about 35 years ago, when he was a 23-year-old philosophy graduate student at Cambridge University with a Harvard physics degree. Lloyd realized that quantum uncertainty, and the way it spreads as particles become increasingly entangled, could replace human uncertainty in the old classical proofs as the true source of the arrow of time.

Using an obscure approach to quantum mechanics that treated units of information as its basic building blocks, Lloyd spent several years studying the evolution of particles in terms of shuffling 1s and 0s. He found that as the particles became increasingly entangled with one another, the information that originally described them (a "1" for clockwise spin and a "0" for counterclockwise, for example) would shift to describe the system of entangled particles as a whole. It was as though the particles gradually lost their individual autonomy and became pawns of the collective state. Eventually, the correlations contained all the information, and the individual particles contained none. At that point, Lloyd discovered, particles arrived at a state of equilibrium, and their states stopped changing, like coffee that has cooled to room temperature.

"What's really going on is things are becoming more correlated with each other," Lloyd recalls realizing. "The arrow of time is an arrow of increasing correlations."

The idea, presented in his 1988 doctoral thesis, fell on deaf ears.[6] When he submitted it to a journal, he was told that there was "no physics in this paper." Quantum information theory "was profoundly unpopular" at the time, Lloyd said, and questions about time's arrow "were for crackpots and Nobel laureates who have gone soft in the head," he remembers one physicist telling him.

"I was darn close to driving a taxicab," Lloyd said.

Advances in quantum computing have since turned quantum information theory into one of the most active branches of physics. Lloyd is now a professor at the Massachusetts Institute of Technology, recognized as one of the founders of the discipline, and his overlooked idea has resurfaced in a stronger form in the hands of the Bristol physicists. The newer proofs are more general, researchers say, and hold for virtually any quantum system.

"When Lloyd proposed the idea in his thesis, the world was not ready," said Renato Renner, head of the Institute for Theoretical Physics at ETH Zurich. "No one understood it. Sometimes you have to have the idea at the right time."

In 2009, the Bristol group's proof resonated with quantum information theorists, opening up new uses for their techniques. It showed that as objects interact with their surroundings—as the particles in a cup of coffee collide with the air, for example—information about their properties "leaks out and becomes smeared over the entire environment," Popescu explained. This local information loss causes the state of the coffee to stagnate even as the pure state of the entire room continues to evolve. Except for rare, random fluctuations, he said, "its state stops changing in time."

Consequently, a tepid cup of coffee does not spontaneously warm up. In principle, as the pure state of the room evolves, the coffee could suddenly become unmixed from the air and enter a pure state of its own. But there are so many more mixed states than pure states available to the coffee that this practically never happens—one would have to outlive the universe to witness it. This statistical unlikelihood gives time's arrow the appearance of irreversibility. "Essentially entanglement opens a very large space for you," Popescu said. "It's like you are at the park and you start next to the gate, far from equilibrium. Then you enter and you have this enormous place and you get lost in it. And you never come back to the gate."

In the new story of the arrow of time, it is the loss of information through quantum entanglement, rather than a subjective lack of human knowledge, that drives a cup of coffee into equilibrium with the surrounding room. The room eventually equilibrates with the outside environment, and the environment drifts even more slowly toward equilibrium with the rest of the universe. The giants of 19th-century thermodynamics viewed this process as a gradual dispersal of energy that increases the overall entropy, or disorder, of the universe. Today, Lloyd, Popescu and others in their field see the arrow of time differently. In their view, information becomes increasingly diffuse, but it never disappears completely. So, they assert, although entropy increases locally, the overall entropy of the universe stays constant at zero.

"The universe as a whole is in a pure state," Lloyd said. "But individual pieces of it, because they are entangled with the rest of the universe, are in mixtures."

One aspect of time's arrow remains unsolved. "There is nothing in these works to say why you started at the gate," Popescu said, referring to the park analogy. "In other words, they don't explain why the initial state of the universe was far from equilibrium." He said this is a question about the nature of the Big Bang.

Despite the recent progress in calculating equilibration time scales, the new approach has yet to make headway as a tool for parsing the thermodynamic properties of specific things, like coffee, glass or exotic states of matter. (Several traditional thermodynamicists reported being only vaguely aware of the new approach.) "The thing is to find the criteria for which things behave like window glass and which things behave like a cup of tea," Renner said. "I would see the new papers as a step in this direction, but much more needs to be done."

Some researchers expressed doubt that this abstract approach to thermodynamics will ever be up to the task of addressing the "hard nitty-gritty of how specific observables behave," as Lloyd put it. But the conceptual advance and new mathematical formalism is already helping researchers address theoretical questions about thermodynamics, such as the fundamental limits of quantum computers and even the ultimate fate of the universe.

"We've been thinking more and more about what we can do with quantum machines," said Paul Skrzypczyk, a quantum physicist now at the University of Bristol. "Given that a system is not yet at equilibrium, we want to get work out of it. How much useful work can we extract? How can I intervene to do something interesting?"[7]

Sean Carroll, a theoretical physicist at Caltech, is employing the new formalism in his latest work on time's arrow in cosmology. "I'm interested in the ultra-long-term fate of cosmological space-times," said Carroll, author of *From Eternity to Here: The Quest for the Ultimate Theory of Time*. "That's a situation where we don't really know all of the relevant laws of physics, so it makes sense to think on a very abstract level, which is why I found this basic quantum-mechanical treatment useful."

Thirty years after Lloyd's big idea about time's arrow fell flat, he is pleased to be witnessing its rise and has been applying the ideas in recent work on the black hole information paradox. "I think now the consensus would be that there is physics in this," he said.

Not to mention a bit of philosophy.

According to the scientists, our ability to remember the past but not the future, another historically confounding manifestation of time's arrow, can also be understood as a buildup of correlations between interacting particles. When you read a message on a piece of paper, your brain becomes correlated with it through the photons that reach your eyes. Only from that moment on will you be capable of remembering what the message says. As Lloyd put it: "The present can be defined by the process of becoming correlated with our surroundings."

The backdrop for the steady growth of entanglement throughout the universe is, of course, time itself. The physicists stress that despite great advances in understanding how changes in time occur, they have made no progress in uncovering the nature of time itself or why it seems different (both perceptually and in the equations of quantum mechanics) than the three dimensions of space. Popescu calls this "one of the greatest unknowns in physics."

"We can discuss the fact that an hour ago, our brains were in a state that was correlated with fewer things," he said. "But our perception that time is flowing—that is a different matter altogether. Most probably, we will need a further revolution in physics that will tell us about that."

QUANTUM WEIRDNESS NOW A MATTER OF TIME

George Musser

I n November 2015, construction workers at the Massachusetts Institute of Technology came across a time capsule 942 years too soon. Buried in 1957 and intended for 2957, the capsule was a glass cylinder filled with inert gas to preserve its contents; it was even laced with carbon-14 so that future researchers could confirm the year of burial, the way they would date a fossil. MIT administrators planned to repair, reseal and rebury it. But is it possible to make it absolutely certain that a message to the future won't be read before its time?

Quantum physics offers a way. In 2012, Jay Olson and Timothy Ralph, both physicists at the University of Queensland in Australia, laid out a procedure to encrypt data so that it can be decrypted only at a specific moment in the future.[1] Their scheme exploits quantum entanglement, a phenomenon in which particles or points in a field, such as the electromagnetic field, shed their separate identities and assume a shared existence, their properties becoming correlated with one another's. Normally physicists think of these correlations as spanning space, linking far-flung locations in a phenomenon that Albert Einstein famously described as "spooky action at a distance." But a growing body of research is investigating how these correlations can span time as well. What happens now can be correlated with what happens later, in ways that elude a simple mechanistic explanation. In effect, you can have spooky action at a delay.

These correlations seriously mess with our intuitions about time and space. Not only can two events be correlated, linking the earlier one to the later one, but two events can become correlated such that it becomes impossible to say which is earlier and which is later. Each of these events is the cause of the other, as if each were the first to occur. (Even a single observer can encounter this causal ambiguity, so it's distinct from the temporal reversals that can happen when two observers move at different velocities, as described in Einstein's special theory of relativity.)

The time-capsule idea is only one demonstration of the potential power of these temporal correlations. They might also boost the speed of quantum computers and strengthen quantum cryptography.

But perhaps most important, researchers hope that the work will open up a new way to unify quantum theory with Einstein's general theory of relativity, which describes the structure of space-time. The world we experience in daily life, in which events occur in an order determined by their locations in space and time, is just a subset of the possibilities that quantum physics allows. "If you have space-time, you have a well-defined causal order," said Časlav Brukner, a physicist at the University of Vienna who studies quantum information. But "if you don't have a well-defined causal order," he said—as is the case in experiments he has proposed—then "you don't have space-time." Some physicists take this as evidence for a profoundly nonintuitive worldview, in which quantum correlations are more fundamental than space-time, and space-time itself is somehow built up from correlations among events, in what might be called quantum relationalism. The argument updates Gottfried Leibniz and Ernst Mach's idea that space-time might not be a God-given backdrop to the world, but instead might derive from the material contents of the universe.

HOW TIME ENTANGLEMENT WORKS

To understand entanglement in time, it helps to first understand entanglement in space, as the two are closely related. In the spatial version of a classic entanglement experiment, two particles, such as photons, are prepared in a shared quantum state, then sent flying in different directions. An observer, Alice, measures the polarization of one photon, and her partner, Bob, measures the other. Alice might measure polarization along the horizontal axis while Bob looks along a diagonal. Or she might choose the vertical angle and he might measure an oblique one. The permutations are endless.

The outcomes of these measurements will match, and what's weird is that they match even when Alice and Bob vary their choice of measurement—as though Alice's particle knew what happened to Bob's, and vice versa. This is true even when nothing connects the particles—no force, wave or carrier pigeon. The correlation appears to violate "locality," the rule that states that effects have causes, and chains of cause and effect must be unbroken in space and time.

In the temporal case, though, the mystery is subtler, involving just a single polarized photon. Alice measures it, and then Bob remeasures it. Distance in space is replaced by an interval of time. The probability of their seeing the same outcome varies with the angle between the polarizers; in fact, it varies in just the same way as in the spatial case. On one level, this does not seem to be strange. Of course what we do first affects what happens next. Of course a particle can communicate with its future self.

The strangeness comes through in an experiment conceived by Robert Spekkens, a physicist who studies the foundations of quantum mechanics at the Perimeter Institute for Theoretical Physics in Waterloo, Canada. Spekkens and his colleagues carried out the experiment in 2009. Alice prepares a photon in one of four possible ways. Classically, we could think of these four ways as two bits of information. Bob then measures the particle in one of two possible ways. If he chooses to measure the particle in the first way, he obtains Alice's first bit of information; if he chooses the second, he obtains her second bit. (Technically, he does not get either bit with certainty, just with a high degree of probability.) The obvious explanation for this result would be if the photon stores both bits and releases one based on Bob's choice. But if that were the case, you'd expect Bob to be able to obtain information about both bits—to measure both of them or at least some characteristic of both, such as whether they are the same or different. But he can't. No experiment, even in principle, can get at both bits—a restriction known as the Holevo bound. "Quantum systems seem to have more memory, but you can't actually access it," said Costantino Budroni, a physicist now at the Institute for Quantum Optics and Quantum Information in Vienna.

The photon really does seem to hold just one bit, and it is as if Bob's choice of measurement retroactively decides which it is. Perhaps that really is what happens, but this is tantamount to time travel—on an oddly limited basis, involving the ability to determine the nature of the bit but denying any glimpse of the future.

Another example of temporal entanglement comes from a team led by Stephen Brierley, a mathematical physicist at the University of Cambridge. In a 2015 paper, Brierley and his collaborators explored the bizarre intersection of entanglement, information and time. If Alice and Bob choose from just two polarizer orientations, the correlations they see are readily explained by a particle carrying a single bit. But if they choose among eight possible directions and they measure and remeasure the particle 16 times, they see correlations that a single bit of memory can't explain. "What we have proven rigorously is that, if you propagate in time the number of bits that corresponds to this Holevo bound, then you definitely cannot explain what quantum mechanics predicts," said Tomasz Paterek, a physicist at Nanyang Technological University in Singapore, and one of Brierley's co-authors. In short, what Alice does to the particle at the beginning of the experiment is correlated with what Bob sees at the end in a way that's too strong to be easily explained. You might call this "supermemory," except that the category of "memory" doesn't seem to capture what's going on.

What exactly is it about quantum physics that goes beyond classical physics to endow particles with this supermemory? Researchers have differing opinions. Some say the key is that quantum measurements inevitably disturb a particle. A disturbance, by definition, is something that affects later measurements. In this case, the disturbance leads to the predicted correlation.

In 2009 Michael Goggin, a physicist who was then at the University of Queensland, and his colleagues did an experiment to get at this issue. They used the trick of spatially entangling a particle with another of its kind and measuring that stand-in particle rather than the original. The measurement of the stand-in still disrupts the original particle (because the two are entangled), but researchers can control the amount that the original is disrupted by varying the degree of entanglement. The trade-off is that the experimenter's knowledge of the original becomes less reliable, but the researchers compensate by testing multiple pairs of particles and aggregating the results in a special way. Goggin and his team reduced the disruption to the point where the original particle was hardly disturbed at all. Measurements at different times were still closely correlated. In fact, they were even more closely correlated than when the measurements disturbed the particle the most. So the question of a particle's supermemory remains a mystery. For now, if you ask why quantum particles produce the strong temporal correlations, physicists basically will answer: "Because."

QUANTUM TIME CAPSULES

Things get more interesting still—offering the potential for quantum time capsules and other fun stuff—when we move to quantum field theory, a more advanced version of quantum mechanics that describes the electromagnetic field and other fields of nature. A field is a highly entangled system. Different parts of it are mutually correlated: A random fluctuation of the field in one place will be matched by a random fluctuation in another. ("Parts" here refers both to regions of space and to spans of time.)

Even a perfect vacuum, which is defined as the absence of particles, will still have quantum fields. And these fields are always vibrating. Space looks empty because the vibrations cancel each other out. And to do this, they must be entangled. The cancellation requires the full set of vibrations; a subset won't necessarily cancel out. But a subset is all you ever see.

If an idealized detector just sits in a vacuum, it will not detect particles. However, any practical detector has a limited range. The field will appear imbalanced to it, and it will detect particles in a vacuum, clicking away like a Geiger counter in a uranium mine. In 1976 Bill Unruh, a theoretical physicist at the University of British Columbia, showed that the detection rate goes up if the detector is accelerating, since the detector loses sensitivity to the regions of space it is moving away from. Accelerate it very strongly and it will click like mad, and the particles it sees will be entangled with particles that remain beyond its view.

In 2011 Olson and Ralph showed that much the same thing happens if the detector can be made to accelerate through time. They described a detector that is sensitive to photons of a single frequency at any one time. The detector sweeps through frequencies like a police radio scanner, moving from lower to higher frequencies (or the other way around). If it sweeps at a quickening pace, it will scan right off the end of the radio dial and cease to function altogether. Because the detector works for only a limited period of time, it lacks sensitivity to the full range of field vibrations, creating the same imbalances that Unruh predicted. Only now, the particles it picks up will be entangled with particles in a hidden region of time—namely, the future.

Olson and Ralph suggest constructing the detector from a loop of superconducting material. Tuned to pick up near-infrared light and completing a scan in a few femtoseconds (10^{-15} seconds), the loop would see the vacuum glowing like a gas at room temperature. No feasible detector accelerating through space could achieve that, so Olson and Ralph's experiment would be an important test of quantum field theory. It could also vindicate Stephen Hawking's ideas about black-hole evaporation, which involve the same basic physics.

If you build two such detectors, one that accelerates and one that decelerates at the same rate, then the particles seen by one detector will be correlated with the particles seen by the other. The first detector might pick up a string of stray particles at random intervals. Minutes or years later, the second detector will pick up another string of stray particles at the same intervals—a spooky recurrence of events. "If you just look at them individually, then they're randomly clicking, but if you get a click in one, then you know that there's going to be a click in the other one if you look at a particular time," Ralph said.

These temporal correlations are the ingredients for that quantum time capsule.[2] The original idea for such a contraption goes back to James Franson, a physicist at the University of Maryland, Baltimore County.[3]

(Franson used spacelike correlations; Olson and Ralph say temporal correlations may make it easier.) You write your message, encode each bit in a photon, and use one of your special detectors to measure those photons along with the background field, thus effectively encrypting your bits. You then store the outcome in the capsule and bury it.

At the designated future time, your descendants measure the field with the paired detector. The two outcomes, together, will reconstitute the original information. "The state is disembodied for the time between [the two measurements], but is encoded somehow in these correlations in the vacuum," Ralph said. Because your descendants must wait for the second detector to be triggered, there's no way to unscramble the message before its time.

The same basic procedure would let you generate entangled particles for use in computation and cryptography. "You could do quantum key distribution without actually sending any quantum signal," Ralph said. "The idea is that you just use the correlations that are already there in the vacuum."

THE NATURE OF SPACE-TIME

These temporal correlations are also challenging physicists' assumptions about the nature of space-time. Whenever two events are correlated and it's not a fluke, there are two explanations: One event causes the other, or some third factor causes both. A background assumption to this logic is that events occur in a given order, dictated by their locations in space and time. Since quantum correlations—certainly the spatial kind, possibly the temporal—are too strong to be explained using one of these two explanations, physicists are revisiting their assumptions. "We cannot really explain these correlations," said Ämin Baumeler, a physicist now at the Institute for Quantum Optics and Quantum Information in Vienna. "There's no mechanism for how these correlations appear. So, they don't really fit into our notion of space-time."

Building on an idea by Lucien Hardy, a theoretical physicist at the Perimeter Institute, Brukner and his colleagues have studied how events might be related to one another without presupposing the existence of space-time.[4] If the setup of one event depends on the outcome of another, you deduce that it occurs later; if the events are completely independent, they must occur far apart in space and time. Such an approach puts spatial and temporal correlations on an equal footing. And it also allows for correlations that are neither spatial nor temporal—meaning that the experiments don't all fit together consistently and there's no way to situate them within space and time.

Brukner's group devised a strange thought experiment that illustrates the idea.[5] Alice and Bob each toss a coin. Each person writes the result of his or her own toss on a piece of paper, along with a guess for the other person's outcome. Each person also sends the paper to the other with this information. They do this a number of times and see how well they do.

Normally the rules of the game are set up so that Alice and Bob do this in a certain sequence. Suppose Alice is first. She can only guess at Bob's outcome (which has yet to occur), but she can send her own result to Bob. Alice's guess as to Bob's flip will be right 50 percent of the time, but he will always get hers right. In the next round, Bob goes first, and the roles are reversed. Overall the success rate will be 75 percent. But if you don't presume they do this in a certain sequence, and if they replace the sheet of paper with a quantum particle, they can succeed 85 percent of the time.

If you try to situate this experiment within space and time, you'll be forced to conclude that it involves a limited degree of time travel, so that the person who goes second can communicate his or her result backward in time to the one who goes first. (The Time Patrol will be relieved that no logical paradoxes can arise: No event can become its own cause.)

Brukner and his colleagues at Vienna have performed a real-world experiment that is similar to this.[6] In the experiment, Alice-and-Bob manipulations were carried out by two optical filters. The researchers beamed a stream of photons at a partially silvered mirror, so that half the photons took one path and half another. (It was impossible to tell, without measuring, which path each individual photon went down; in a sense, it took both paths at once.) On the first path, the photons passed through Alice's filter first, followed by Bob's. On the second path, the photons navigated them in reverse order. The experiment took quantum indeterminacy to a whole new level. Not only did the particles not possess definite properties in advance of measurement, the operations performed on them were not even conducted in a definite sequence.

On a practical level, the experiment opens up new possibilities for quantum computers.[7] The filters corresponding to Alice and Bob represent two different mathematical operations, and the apparatus was able to ascertain in a single step whether the order of those operations matters—whether A followed by B is the same as B followed by A. Normally you'd need two steps to do that, so the procedure is a significant speedup. Quantum computers are sometimes described as performing a series of operations on all possible data at once, but they might also be able to perform all possible operations at once.

Now imagine taking this experiment a step further. In Brukner's original experiment, the path of each individual photon is placed into a "superposition"—the photon goes down a quantum combination of the Alice-first path and the Bob-first path. There is no definite answer to the question, "Which filter did the photon go through first?"—until a measurement is carried out and the ambiguity is resolved. If, instead of a photon, a gravitating object could be put into such a temporal superposition, the apparatus would put space-time itself into a superposition. In such a case, the sequence of Alice and Bob would remain ambiguous. Cause and effect would blur together, and you would be unable to give a step-by-step account of what happened.

Only when these indeterminate causal relations between events are pruned away—so that nature realizes only some of the possibilities available to it—do space and time become meaningful. Quantum correlations come first, space-time later. Exactly how does space-time emerge out of the quantum world? Brukner said he is still unsure. As with the time capsule, the answer will come only when the time is right.

A DEBATE OVER THE PHYSICS OF TIME

Dan Falk

E instein once described his friend Michele Besso as "the best sounding board in Europe" for scientific ideas. They attended university together in Zurich; later they were colleagues at the patent office in Bern. When Besso died in the spring of 1955, Einstein—knowing that his own time was also running out—wrote a now-famous letter to Besso's family. "Now he has departed this strange world a little ahead of me," Einstein wrote of his friend's passing. "That signifies nothing. For us believing physicists, the distinction between past, present and future is only a stubbornly persistent illusion."

Einstein's statement was not merely an attempt at consolation. Many physicists argue that Einstein's position is implied by the two pillars of modern physics: Einstein's masterpiece, the general theory of relativity, and the Standard Model of particle physics. The laws that underlie these theories are time-symmetric—that is, the physics they describe is the same, regardless of whether the variable called "time" increases or decreases. Moreover, they say nothing at all about the point we call "now"—a special moment (or so it appears) for us, but seemingly undefined when we talk about the universe at large. The resulting timeless cosmos is sometimes called a "block universe"—a static block of space-time in which any flow of time, or passage through it, must presumably be a mental construct or other illusion.

Many physicists have made peace with the idea of a block universe, arguing that the task of the physicist is to describe how the universe appears from the point of view of individual observers. To understand the distinction between past, present and future, you have to "plunge into this block universe and ask: 'How is an observer perceiving time?'" said Andreas Albrecht, a physicist at the University of California, Davis, and one of the founders of the theory of cosmic inflation.

Others vehemently disagree, arguing that the task of physics is to explain not just how time appears to pass, but why. For them, the universe is not static. The passage of time is physical. "I'm sick and tired of this block universe," said Avshalom Elitzur, a physicist and philosopher formerly of Bar-Ilan University. "I don't think that next Thursday has the same footing as this Thursday. The future does not exist. It does not! Ontologically, it's not there."

In June 2016, about 60 physicists, along with a handful of philosophers and researchers from other branches of science, gathered at the Perimeter Institute for Theoretical Physics in Waterloo, Canada, to debate this question at the Time in Cosmology conference. The conference was co-organized by the physicist Lee Smolin, an outspoken critic of the block-universe idea. His position is spelled out for a lay audience in *Time Reborn* and in a more technical work, *The Singular Universe and the Reality of Time*, co-authored with the philosopher Roberto Mangabeira Unger, who was also a co-organizer of the conference. In the latter work, mirroring Elitzur's sentiments about the future's lack of concreteness, Smolin wrote: "The future is not now real and there can be no definite facts of the matter about the future." What is real is "the process by which future events are generated out of present events," he said at the conference.

Those in attendance wrestled with several questions: the distinction between past, present and future; why time appears to move in only one direction; and whether time is fundamental or emergent. Most of those issues, not surprisingly, remained unresolved. But for four days, participants listened attentively to the latest proposals for tackling these questions— and, especially, to the ways in which we might reconcile our perception of time's passage with a static, seemingly timeless universe.

TIME SWEPT UNDER THE RUG

There are a few things that everyone agrees on. The directionality that we observe in the macroscopic world is very real: Teacups shatter but do not spontaneously reassemble; eggs can be scrambled but not unscrambled. Entropy—a measure of the disorder in a system—always increases, a fact encoded in the second law of thermodynamics. As the Austrian physicist Ludwig Boltzmann understood in the 19th century, the second law explains why events are more likely to evolve in one direction rather than another. It accounts for the arrow of time.

But things get trickier when we step back and ask why we happen to live in a universe where such a law holds. "What Boltzmann truly explained is why the entropy of the universe will be larger tomorrow than it is today," said Sean Carroll of Caltech, as we sat in a hotel bar after the second day of presentations. "But if that was all you knew, you'd also say that the entropy of the universe was probably larger yesterday than today—because all the underlying dynamics are completely symmetric with respect to time." That is, if entropy is ultimately based on the underlying laws of the universe, and those laws are the same going forward and backward, then entropy is just as likely to increase going backward in time. But no one believes that entropy actually works that way. Scrambled eggs always come after whole eggs, never the other way around.

To make sense of this, physicists have proposed that the universe began in a very special low-entropy state. In this view, which the Columbia University philosopher of physics David Albert named the "past hypothesis," entropy increases because the Big Bang happened to produce an exceptionally low-entropy universe. There was nowhere to go but up. The past hypothesis implies that every time we cook an egg, we're taking advantage of events that happened nearly 14 billion years ago. "What you need the Big Bang to explain is: 'Why were there ever unbroken eggs?'" Carroll said.

Some physicists are more troubled than others by the past hypothesis. Taking things we don't understand about the physics of today's universe and saying the answer can be found in the Big Bang could be seen, perhaps, as passing the buck—or as sweeping our problems under the carpet. Every time we invoke initial conditions, "the pile of things under the rug gets bigger," said Marina Cortes, a cosmologist at the Royal Observatory in Edinburgh and a co-organizer of the conference.

To Smolin, the past hypothesis feels more like an admission of failure than a useful step forward. As he puts it in *The Singular Universe*: "The fact to be explained is why the universe, even 13.8 billion years after the Big Bang, has not reached equilibrium, which is by definition the most probable state, and it hardly suffices to explain this by asserting that the universe started in an even less probable state than the present one."

Other physicists, however, point out that it's normal to develop theories that can describe a system given certain initial conditions. A theory needn't strive to explain those conditions.

Another set of physicists think that the past hypothesis, while better than nothing, is more likely to be a placeholder than a final answer. Perhaps, if we're lucky, it will point the way to something deeper. "Many people say that the past hypothesis is just a fact, and there isn't any underlying way to explain it. I don't rule out that possibility," Carroll said. "To me, the past hypothesis is a clue to help us develop a more comprehensive view of the universe."

THE ALTERNATIVE ORIGINS OF TIME

Can the arrow of time be understood without invoking the past hypothesis? Some physicists argue that gravity—not thermodynamics—aims time's arrow. In this view, gravity causes matter to clump together, defining an arrow of time that aligns itself with growth of complexity, said Tim Koslowski, a physicist at the National Autonomous University of Mexico (he described the idea in a 2014 paper co-authored by the British physicist Julian Barbour and Flavio Mercati, a physicist at Perimeter).[1] Koslowski and his colleagues developed simple models of universes made up of 1,000 pointlike particles, subject only to Newton's law of gravitation, and found that there will always be a moment of maximum density and minimum complexity. As one moves away from that point, in either direction, complexity increases. Naturally, we—complex creatures capable of making observations—can only evolve at some distance from the minimum. Still, wherever we happen to find ourselves in the history of the universe, we can point to an era of less complexity and call it the past, Koslowski said. The models are globally time-symmetric, but every observer will experience a local arrow of time. It's significant that the low-entropy starting point isn't an add-on to the model. Rather, it emerges naturally from it. "Gravity essentially eliminates the need for a past hypothesis," Koslowski said.

The idea that time moves in more than one direction, and that we just happen to inhabit a section of the cosmos with a single, locally defined arrow of time, isn't new. Back in 2004, Carroll, along with his graduate student Jennifer Chen, put forward a similar proposal based on eternal inflation, a relatively well-known model of the beginning of the universe.[2] Carroll sees the work of Koslowski and his colleagues as a useful step, especially since they worked out the mathematical details of their model (he and Chen did not). Still, he has some concerns. For example, he said it's not clear that gravity plays as important a role as their paper claims. "If you just had particles in empty space, you'd get exactly the same qualitative behavior," he said.

Increasing complexity, Koslowski said, has one crucial side effect: It leads to the formation of certain arrangements of matter that maintain their structure over time. These structures can store information; Koslowski calls them "records." Gravity is the first and primary force that makes record formation possible; other processes then give rise to everything from fossils and tree rings to written documents. What all of these entities have in common is that they contain information about some earlier state of the universe. I asked Koslowski if memories stored in brains are another kind of record. Yes, he said. "Ideally we would be able to build ever more complex models, and come eventually to the memory in my phone, the memory in my brain, in history books." A more complex universe contains more records than a less complex universe, and this, Koslowski said, is why we remember the past but not the future.

But perhaps time is even more fundamental than this. For George Ellis, a cosmologist at the University of Cape Town in South Africa, time is a more basic entity, one that can be understood by picturing the block universe as itself evolving. In his "evolving block universe" model, the universe is a growing volume of space-time.[3] The surface of this volume can be thought of as the present moment. The surface represents the instant where "the indefiniteness of the future changes to the definiteness of the past," as he described it. "Space-time itself is growing as time passes." One can discern the direction of time by looking at which part of the universe is fixed (the past) and which is changing (the future). Although some colleagues disagree, Ellis stresses that the model is a modification, not a radical overhaul, of the standard view. "This is a block universe with dynamics covered by the general-relativity field equations—absolutely standard—but with a future boundary that is the ever-changing present," he said. In this view, while the past is fixed and unchangeable, the future is open. The model "obviously represents the passing of time in a more satisfactory way than the usual block universe," he said.

Unlike the traditional block view, Ellis's picture appears to describe a universe with an open future—seemingly in conflict with a law-governed universe in which past physical states dictate future states. (Although quantum uncertainty, as Ellis pointed out, may be enough to sink such a deterministic view.) At the conference, someone asked Ellis if, given enough information about the physics of a sphere of a certain radius centered on the British Midlands in early June 2016, one could have predicted the result of the Brexit vote. "Not using physics," Ellis replied. For that, he said, we'd need a better understanding of how minds work.

Another approach that aims to reconcile the apparent passage of time with the block universe goes by the name of causal set theory. First developed in the 1980s as an approach to quantum gravity by the physicist Rafael Sorkin—who was also at the conference—the theory is based on the idea that space-time is discrete rather than continuous. In this view, although the universe appears continuous at the macroscopic level, if we could peer down to the so-called Planck scale (distances of about 10^{-35} meters) we'd discover that the universe is made up of elementary units or "atoms" of space-time. The atoms form what mathematicians call a "partially ordered set"—an array in which each element is linked to an adjacent element in a particular sequence. The number of these atoms (estimated to be a whopping 10^{240} in the visible universe) gives rise to the volume of space-time, while their sequence gives rise to time. According to the theory, new space-time atoms are continuously coming into existence. Fay Dowker, a physicist at Imperial College London, referred to this at the conference as "accretive time." She invited everyone to think of space-time as accreting new space-time atoms in a way roughly analogous to a seabed depositing new layers of sediment over time. General relativity yields only a block, but causal sets seem to allow a "becoming," she said. "The block universe is a static thing—a static picture of the world—whereas this process of becoming is dynamical." In this view, the passage of time is a fundamental rather than an emergent feature of the cosmos. (Causal set theory has made at least one successful prediction about the universe, Dowker pointed out, having been used to estimate the value of the cosmological constant based only on the space-time volume of the universe.[4])

THE PROBLEM WITH THE FUTURE

In the face of these competing models, many thinkers seem to have stopped worrying and learned to love (or at least tolerate) the block universe.

Perhaps the strongest statement made at the conference in favor of the block universe's compatibility with everyday experience came from the philosopher Jenann Ismael of the University of Arizona. The way Ismael sees it, the block universe, properly understood, holds within it the explanation for our experience of time's apparent passage. A careful look at conventional physics, supplemented by what we've learned in recent decades from cognitive science and psychology, can recover "the flow, the whoosh, of experience," she said. In this view, time is not an illusion—in fact, we experience it directly. She cited studies that show that each moment we experience represents a finite interval of time. In other words, we don't infer the flow of time; it's part of the experience itself. The challenge, she said, is to frame this first-person experience within the static block offered by physics— to examine "how the world looks from the evolving frame of reference of an embedded perceiver" whose history is represented by a curve within the space-time of the block universe.

Ismael's presentation drew a mixed response. Carroll said he agreed with everything she had said; Elitzur said he "wanted to scream" during her talk. (He later clarified: "If I bang my head against the wall, it's because I hate the future.") An objection voiced many times during the conference was that the block universe seems to imply, in some important way, that the future already exists, yet statements about, say, next Thursday's weather are neither true nor false. For some, this seems like an insurmountable problem with the block-universe view. Ismael had heard these objections many times before. Future events exist, she said, they just don't exist *now*. "The block universe is not a changing picture," she said. "It's a picture of change." Things happen when they happen. "This is a moment—and I know everybody here is going to hate this—but physics could do with some philosophy," she said. "There's a long history of discussion about the truth-values of future contingent statements—and it really has nothing to do with the experience of time." And for those who wanted to read more? "I recommend Aristotle," she said.

IV WHAT IS LIFE?

A NEW PHYSICS THEORY OF LIFE

Natalie Wolchover

W hy does life exist?
Popular hypotheses credit a primordial soup, a bolt of lightning and a colossal stroke of luck. But if a provocative new theory is correct, luck may have little to do with it. Instead, according to the physicist proposing the idea, the origin and subsequent evolution of life follow from the fundamental laws of nature and "should be as unsurprising as rocks rolling downhill."

From the standpoint of physics, there is one essential difference between living things and inanimate clumps of carbon atoms: The former tend to be much better at capturing energy from their environment and dissipating that energy as heat. Jeremy England, a 37-year-old associate professor at MIT, derived in 2013 a mathematical formula that he believes explains this capacity. The formula, based on established physics, indicates that when a group of atoms is driven by an external source of energy (like the sun or chemical fuel) and surrounded by a heat bath (like the ocean or atmosphere), it will often gradually restructure itself in order to dissipate increasingly more energy. This could mean that under certain conditions, matter inexorably acquires the key physical attribute associated with life.

"You start with a random clump of atoms, and if you shine light on it for long enough, it should not be so surprising that you get a plant," England said.

England's theory is meant to underlie, rather than replace, Darwin's theory of evolution by natural selection, which provides a powerful description of life at the level of genes and populations. "I am certainly not saying that Darwinian ideas are wrong," he explained. "On the contrary, I am just saying that from the perspective of the physics, you might call Darwinian evolution a special case of a more general phenomenon."

His idea has sparked controversy among his colleagues, who see it as either tenuous or a potential breakthrough, or both.[1]

England has taken "a very brave and very important step," said Alexander Grosberg, a professor of physics at New York University who has followed England's work since its early stages. The "big hope" is that he has identified the underlying physical principle driving the origin and evolution of life, Grosberg said.

"Jeremy is just about the brightest young scientist I ever came across," said Attila Szabo, a biophysicist in the Laboratory of Chemical Physics at the National Institutes of Health who corresponded with England about his theory after meeting him at a conference. "I was struck by the originality of the ideas."

Others, such as Eugene Shakhnovich, a professor of chemistry, chemical biology and biophysics at Harvard University, are not convinced. "Jeremy's ideas are interesting and potentially promising, but at this point are extremely speculative, especially as applied to life phenomena," Shakhnovich said.

England's theoretical results are generally considered valid. It is his interpretation—that his formula represents the driving force behind a class of phenomena in nature that includes life—that remains unproven. But already, there are ideas about how to test that interpretation in the lab.

"He's trying something radically different," said Mara Prentiss, a professor of physics at Harvard who is contemplating such an experiment after learning about England's work. "As an organizing lens, I think he has a fabulous idea. Right or wrong, it's going to be very much worth the investigation."

At the heart of England's idea is the second law of thermodynamics, also known as the law of increasing entropy or the "arrow of time." Energy tends to disperse or spread out as time progresses. It increases as a simple matter of probability: There are more ways for energy to be spread out than for it to be concentrated. Eventually, the system arrives at a state of maximum entropy called "thermodynamic equilibrium," in which energy is uniformly distributed.

Although entropy must increase over time in an isolated or "closed" system, an "open" system can keep its entropy low—that is, divide energy unevenly among its atoms—by greatly increasing the entropy of its surroundings. In his influential 1944 monograph *What Is Life?* the eminent quantum physicist Erwin Schrödinger argued that this is what living things must do. A plant, for example, absorbs extremely energetic sunlight, uses it to build sugars and ejects infrared light, a much less concentrated form of energy. The overall entropy of the universe increases during photosynthesis as the sunlight dissipates, even as the plant prevents itself from decaying by maintaining an orderly internal structure.

Life does not violate the second law of thermodynamics, but until recently, physicists were unable to use thermodynamics to explain why it should arise in the first place. In Schrödinger's day, they could solve the equations of thermodynamics only for closed systems in equilibrium. In the 1960s, the Belgian physicist Ilya Prigogine made progress on predicting the behavior of open systems weakly driven by external energy sources (for which he won the 1977 Nobel Prize in chemistry). But the behavior of systems that are far from equilibrium, which are connected to the outside environment and strongly driven by external sources of energy, could not be predicted.

This situation changed in the late 1990s, due primarily to the work of Chris Jarzynski, now at the University of Maryland, and Gavin Crooks, now at Lawrence Berkeley National Laboratory. Jarzynski and Crooks showed that the entropy produced by a thermodynamic process, such as the cooling of a cup of coffee, corresponds to a simple ratio: the probability that the atoms will undergo that process divided by their probability of undergoing the reverse process (that is, spontaneously interacting in such a way that the coffee warms up).[2] As entropy production increases, so does this ratio: A system's behavior becomes more and more "irreversible." The simple yet rigorous formula could in principle be applied to any thermodynamic process, no matter how fast or far from equilibrium. "Our understanding of far-from-equilibrium statistical mechanics greatly improved," Grosberg said. England, who is trained in both biochemistry and physics, started his own lab at MIT six years ago and decided to apply the new knowledge of statistical physics to biology.

Using Jarzynski and Crooks' formulation, he derived a generalization of the second law of thermodynamics that holds for systems of particles with certain characteristics: The systems are strongly driven by an external energy source such as an electromagnetic wave, and they can dump heat into a surrounding bath. This class of systems includes all living things. England then determined how such systems tend to evolve over time as they increase their irreversibility. "We can show very simply from the formula that the more likely evolutionary outcomes are going to be the ones that absorbed and dissipated more energy from the environment's external drives on the way to getting there," he said. The finding makes intuitive sense: Particles tend to dissipate more energy when they resonate with a driving force, or move in the direction it is pushing them, and they are more likely to move in that direction than any other at any given moment.

"This means clumps of atoms surrounded by a bath at some temperature, like the atmosphere or the ocean, should tend over time to arrange themselves to resonate better and better with the sources of mechanical, electromagnetic or chemical work in their environments," England explained.

Self-replication (or reproduction, in biological terms), the process that drives the evolution of life on Earth, is one such mechanism by which a system might dissipate an increasing amount of energy over time. As England put it, "A great way of dissipating more is to make more copies of yourself." In a September 2013 paper in the *Journal of Chemical Physics*, he reported the theoretical minimum amount of dissipation that can occur during the self-replication of RNA molecules and bacterial cells, and showed that it is very close to the actual amounts these systems dissipate when replicating.[3] He also showed that RNA, the nucleic acid that many scientists believe served as the precursor to DNA-based life, is a particularly cheap building material. Once RNA arose, he argues, its "Darwinian takeover" was perhaps not surprising.

The chemistry of the primordial soup, random mutations, geography, catastrophic events and countless other factors have contributed to the fine details of Earth's diverse flora and fauna. But according to England's theory, the underlying principle driving the whole process is dissipation-driven adaptation of matter.

This principle would apply to inanimate matter as well. "It is very tempting to speculate about what phenomena in nature we can now fit under this big tent of dissipation-driven adaptive organization," England said. "Many examples could just be right under our nose, but because we haven't been looking for them we haven't noticed them."

Scientists have already observed self-replication in nonliving systems. According to new research led by Philip Marcus of the University of California, Berkeley, and reported in *Physical Review Letters* in August 2013, vortices in turbulent fluids spontaneously replicate themselves by drawing energy from shear in the surrounding fluid.[4] And in a January 2014 paper in *Proceedings of the National Academy of Sciences*, Michael Brenner, a professor of applied mathematics and physics at Harvard, and his collaborators presented theoretical models and simulations of microstructures that self-replicate.[5] These clusters of specially coated microspheres dissipate energy by roping nearby spheres into forming identical clusters. "This connects very much to what Jeremy is saying," Brenner said.

Besides self-replication, greater structural organization is another means by which strongly driven systems ramp up their ability to dissipate energy. A plant, for example, is much better at capturing and routing solar energy through itself than an unstructured heap of carbon atoms. Thus, England argues that under certain conditions, matter will spontaneously self-organize. This tendency could account for the internal order of living things and of many inanimate structures as well. "Snowflakes, sand dunes and turbulent vortices all have in common that they are strikingly patterned structures that emerge in many-particle systems driven by some dissipative process," he said. Condensation, wind and viscous drag are the relevant processes in these particular cases.

"He is making me think that the distinction between living and non-living matter is not sharp," said Carl Franck, a biological physicist at Cornell University, in an email. "I'm particularly impressed by this notion when one considers systems as small as chemical circuits involving a few biomolecules."

England's bold idea will likely face close scrutiny in the coming years. He is currently running computer simulations to test his theory that systems of particles adapt their structures to become better at dissipating energy. The next step will be to run experiments on living systems.

Prentiss, who runs an experimental biophysics lab at Harvard, says England's theory could be tested by comparing cells with different mutations and looking for a correlation between the amount of energy the cells dissipate and their replication rates. "One has to be careful because any mutation might do many things," she said. "But if one kept doing many of these experiments on different systems and if [dissipation and replication success] are indeed correlated, that would suggest this is the correct organizing principle."

Brenner said he hopes to connect England's theory to his own microsphere constructions and determine whether the theory correctly predicts which self-replication and self-assembly processes can occur—"a fundamental question in science," he said.

Having an overarching principle of life and evolution would give researchers a broader perspective on the emergence of structure and function in living things, many of the researchers said. "Natural selection doesn't explain certain characteristics," said Ard Louis, a biophysicist at Oxford University, in an email. These characteristics include a heritable change to gene expression called methylation, increases in complexity in the absence of natural selection and certain molecular changes Louis has recently studied.

If England's approach stands up to more testing (his recent computer simulations appeared to support his general thesis, but the implications for real life remain speculative), it could further liberate biologists from seeking a Darwinian explanation for every adaptation and allow them to think more generally in terms of dissipation-driven organization. They might find, for example, that "the reason that an organism shows characteristic X rather than Y may not be because X is more fit than Y, but because physical constraints make it easier for X to evolve than for Y to evolve," Louis said.

"People often get stuck in thinking about individual problems," Prentiss said. Whether or not England's ideas turn out to be exactly right, she said, "thinking more broadly is where many scientific breakthroughs are made."

HOW LIFE (AND DEATH) SPRING FROM DISORDER

Philip Ball

W hat's the difference between physics and biology? Take a golf ball and a cannonball and drop them off the Tower of Pisa. The laws of physics allow you to predict their trajectories pretty much as accurately as you could wish for.

Now do the same experiment again, but replace the cannonball with a pigeon.

Biological systems don't defy physical laws, of course—but neither do they seem to be predicted by them. In contrast, they are goal-directed: survive and reproduce. We can say that they have a purpose—or what philosophers have traditionally called a teleology—that guides their behavior.

By the same token, physics now lets us predict, starting from the state of the universe a billionth of a second after the Big Bang, what it looks like today. But no one imagines that the appearance of the first primitive cells on Earth led predictably to the human race. Laws do not, it seems, dictate the course of evolution.

The teleology and historical contingency of biology, said the evolutionary biologist Ernst Mayr, make it unique among the sciences. Both of these features stem from perhaps biology's only general guiding principle: evolution. It depends on chance and randomness, but natural selection gives it the appearance of intention and purpose. Animals are drawn to water not by some magnetic attraction, but because of their instinct, their intention, to survive. Legs serve the purpose of, among other things, taking us to the water.

Mayr claimed that these features make biology exceptional—a law unto itself. But recent developments in nonequilibrium physics, complex systems science and information theory are challenging that view.

Once we regard living things as agents performing a computation— collecting and storing information about an unpredictable environment— capacities and considerations such as replication, adaptation, agency, purpose and meaning can be understood as arising not from evolutionary improvisation, but as inevitable corollaries of physical laws. In other words, there appears to be a kind of physics of things doing stuff, and evolving to do stuff. Meaning and intention—thought to be the defining characteristics of living systems—may then emerge naturally through the laws of thermodynamics and statistical mechanics.

In November 2016, physicists, mathematicians and computer scientists came together with evolutionary and molecular biologists to talk—and sometimes argue—about these ideas at a workshop at the Santa Fe Institute in New Mexico, the mecca for the science of "complex systems." They asked: Just how special (or not) is biology?

It's hardly surprising that there was no consensus. But one message that emerged very clearly was that, if there's a kind of physics behind biological teleology and agency, it has something to do with the same concept that seems to have become installed at the heart of fundamental physics itself: information.

DISORDER AND DEMONS

The first attempt to bring information and intention into the laws of thermodynamics came in the middle of the 19th century, when statistical mechanics was being invented by the Scottish scientist James Clerk Maxwell. Maxwell showed how introducing these two ingredients seemed to make it possible to do things that thermodynamics proclaimed impossible.

Maxwell had already shown how the predictable and reliable mathematical relationships between the properties of a gas—pressure, volume and temperature—could be derived from the random and unknowable motions of countless molecules jiggling frantically with thermal energy. In other words, thermodynamics—the new science of heat flow, which united large-scale properties of matter like pressure and temperature—was the outcome of statistical mechanics on the microscopic scale of molecules and atoms.

According to thermodynamics, the capacity to extract useful work from the energy resources of the universe is always diminishing. Pockets of energy are declining, concentrations of heat are being smoothed away. In every physical process, some energy is inevitably dissipated as useless heat, lost among the random motions of molecules. This randomness is equated with the thermodynamic quantity called entropy—a measurement of disorder—which is always increasing. That is the second law of thermodynamics. Eventually all the universe will be reduced to a uniform, boring jumble: a state of equilibrium, wherein entropy is maximized and nothing meaningful will ever happen again.

Are we really doomed to that dreary fate? Maxwell was reluctant to believe it, and in 1867 he set out to, as he put it, "pick a hole" in the second law. His aim was to start with a disordered box of randomly jiggling molecules, then separate the fast molecules from the slow ones, reducing entropy in the process.

Imagine some little creature—the physicist William Thomson later called it, rather to Maxwell's dismay, a demon—that can see each individual molecule in the box. The demon separates the box into two compartments, with a sliding door in the wall between them. Every time he sees a particularly energetic molecule approaching the door from the right-hand compartment, he opens it to let it through. And every time a slow, "cold" molecule approaches from the left, he lets that through, too. Eventually, he has a compartment of cold gas on the right and hot gas on the left: a heat reservoir that can be tapped to do work.

This is only possible for two reasons. First, the demon has more information than we do: It can see all of the molecules individually, rather than just statistical averages. And second, it has intention: a plan to separate the hot from the cold. By exploiting its knowledge with intent, it can defy the laws of thermodynamics.

At least, so it seemed. It took a hundred years to understand why Maxwell's demon can't in fact defeat the second law and avert the inexorable slide toward deathly, universal equilibrium. And the reason shows that there is a deep connection between thermodynamics and the processing of information—or in other words, computation. The German-American physicist Rolf Landauer showed that even if the demon can gather information and move the (frictionless) door at no energy cost, a penalty must eventually be paid.[1] Because it can't have unlimited memory of every molecular motion, it must occasionally wipe its memory clean—forget what it has seen and start again—before it can continue harvesting energy. This act of information erasure has an unavoidable price: It dissipates energy, and therefore increases entropy. All the gains against the second law made by the demon's nifty handiwork are canceled by "Landauer's limit": the finite cost of information erasure (or more generally, of converting information from one form to another).

Living organisms seem rather like Maxwell's demon. Whereas a beaker full of reacting chemicals will eventually expend its energy and fall into boring stasis and equilibrium, living systems have collectively been avoiding the lifeless equilibrium state since the origin of life about three and a half billion years ago. They harvest energy from their surroundings to sustain this nonequilibrium state, and they do it with "intention." Even simple bacteria move with "purpose" toward sources of heat and nutrition. In his 1944 book *What is Life?*, the physicist Erwin Schrödinger expressed this by saying that living organisms feed on "negative entropy."

They achieve it, Schrödinger said, by capturing and storing information. Some of that information is encoded in their genes and passed on from one generation to the next: a set of instructions for reaping negative entropy. Schrödinger didn't know where the information is kept or how it is encoded, but his intuition that it is written into what he called an "aperiodic crystal" inspired Francis Crick, himself trained as a physicist, and James Watson when in 1953 they figured out how genetic information can be encoded in the molecular structure of the DNA molecule.

A genome, then, is at least in part a record of the useful knowledge that has enabled an organism's ancestors—right back to the distant past— to survive on our planet. According to David Wolpert, a mathematician and physicist at the Santa Fe Institute who convened the 2016 workshop, and his colleague Artemy Kolchinsky, the key point is that well-adapted organisms are correlated with that environment. If a bacterium swims dependably toward the left or the right when there is a food source in that direction, it is better adapted, and will flourish more, than one that swims in random directions and so only finds the food by chance. A correlation between the state of the organism and that of its environment implies that they share information in common. Wolpert and Kolchinsky say that it's this information that helps the organism stay out of equilibrium—because, like Maxwell's demon, it can then tailor its behavior to extract work from fluctuations in its surroundings. If it did not acquire this information, the organism would gradually revert to equilibrium: It would die.

Looked at this way, life can be considered as a computation that aims to optimize the storage and use of meaningful information.[2] And life turns out to be extremely good at it. Landauer's resolution of the conundrum of Maxwell's demon set an absolute lower limit on the amount of energy a finite-memory computation requires: namely, the energetic cost of forgetting. The best computers today are far, far more wasteful of energy than that, typically consuming and dissipating more than a million times more. But according to Wolpert, "a very conservative estimate of the thermodynamic efficiency of the total computation done by a cell is that it is only 10 or so times more than the Landauer limit."

The implication, he said, is that "natural selection has been hugely concerned with minimizing the thermodynamic cost of computation. It will do all it can to reduce the total amount of computation a cell must perform." In other words, biology (possibly excepting ourselves) seems to take great care not to overthink the problem of survival. This issue of the costs and benefits of computing one's way through life, he said, has been largely overlooked in biology so far.

INANIMATE DARWINISM

So living organisms can be regarded as entities that attune to their environment by using information to harvest energy and evade equilibrium. Sure, it's a bit of a mouthful. But notice that it said nothing about genes and evolution, on which Mayr, like many biologists, assumed that biological intention and purpose depend.

How far can this picture then take us? Genes honed by natural selection are undoubtedly central to biology. But could it be that evolution by natural selection is itself just a particular case of a more general imperative toward function and apparent purpose that exists in the purely physical universe? It is starting to look that way.

Adaptation has long been seen as the hallmark of Darwinian evolution. But Jeremy England at the Massachusetts Institute of Technology has argued that adaptation to the environment can happen even in complex nonliving systems.

Adaptation here has a more specific meaning than the usual Darwinian picture of an organism well-equipped for survival. One difficulty with the Darwinian view is that there's no way of defining a well-adapted organism except in retrospect. The "fittest" are those that turned out to be better at survival and replication, but you can't predict what fitness entails. Whales and plankton are well-adapted to marine life, but in ways that bear little obvious relation to one another.

England's definition of "adaptation" is closer to Schrödinger's, and indeed to Maxwell's: A well-adapted entity can absorb energy efficiently from an unpredictable, fluctuating environment. It is like the person who keeps her footing on a pitching ship while others fall over because she's better at adjusting to the fluctuations of the deck. Using the concepts and methods of statistical mechanics in a nonequilibrium setting, England and his colleagues argue that these well-adapted systems are the ones that absorb and dissipate the energy of the environment, generating entropy in the process.[3]

Complex systems tend to settle into these well-adapted states with surprising ease, said England: "Thermally fluctuating matter often gets spontaneously beaten into shapes that are good at absorbing work from the time-varying environment."

There is nothing in this process that involves the gradual accommodation to the surroundings through the Darwinian mechanisms of replication, mutation and inheritance of traits. There's no replication at all. "What is exciting about this is that it means that when we give a physical account of the origins of some of the adapted-looking structures we see, they don't necessarily have to have had parents in the usual biological sense," said England. "You can explain evolutionary adaptation using thermodynamics, even in intriguing cases where there are no self-replicators and Darwinian logic breaks down"—so long as the system in question is complex, versatile and sensitive enough to respond to fluctuations in its environment.

But neither is there any conflict between physical and Darwinian adaptation. In fact, the latter can be seen as a particular case of the former. If replication is present, then natural selection becomes the route by which systems acquire the ability to absorb work—Schrödinger's negative entropy—from the environment. Self-replication is, in fact, an especially good mechanism for stabilizing complex systems, and so it's no surprise that this is what biology uses. But in the nonliving world where replication doesn't usually happen, the well-adapted dissipative structures tend to be ones that are highly organized, like sand ripples and dunes crystallizing from the random dance of windblown sand. Looked at this way, Darwinian evolution can be regarded as a specific instance of a more general physical principle governing nonequilibrium systems.

PREDICTION MACHINES

This picture of complex structures adapting to a fluctuating environment allows us also to deduce something about how these structures store information. In short, so long as such structures—whether living or not—are compelled to use the available energy efficiently, they are likely to become "prediction machines."

It's almost a defining characteristic of life that biological systems change their state in response to some driving signal from the environment. Something happens; you respond. Plants grow toward the light; they produce toxins in response to pathogens. These environmental signals are typically unpredictable, but living systems learn from experience, storing up information about their environment and using it to guide future behavior. (Genes, in this picture, just give you the basic, general-purpose essentials.)

Prediction isn't optional, though. According to the work of Susanne Still at the University of Hawaii, Gavin Crooks, formerly at the Lawrence Berkeley

National Laboratory in California, and their colleagues, predicting the future seems to be essential for any energy-efficient system in a random, fluctuating environment.[4]

There's a thermodynamic cost to storing information about the past that has no predictive value for the future, Still and colleagues show. To be maximally efficient, a system has to be selective. If it indiscriminately remembers everything that happened, it incurs a large energy cost. On the other hand, if it doesn't bother storing any information about its environment at all, it will be constantly struggling to cope with the unexpected. "A thermodynamically optimal machine must balance memory against prediction by minimizing its nostalgia—the useless information about the past," said a co-author, David Sivak, now at Simon Fraser University in Burnaby, British Columbia. In short, it must become good at harvesting meaningful information—that which is likely to be useful for future survival.

You'd expect natural selection to favor organisms that use energy efficiently. But even individual biomolecular devices like the pumps and motors in our cells should, in some important way, learn from the past to anticipate the future. To acquire their remarkable efficiency, Still said, these devices must "implicitly construct concise representations of the world they have encountered so far, enabling them to anticipate what's to come."

THE THERMODYNAMICS OF DEATH

Even if some of these basic information-processing features of living systems are already prompted, in the absence of evolution or replication, by nonequilibrium thermodynamics, you might imagine that more complex traits—tool use, say, or social cooperation—must be supplied by evolution.

Well, don't count on it. These behaviors, commonly thought to be the exclusive domain of the highly advanced evolutionary niche that includes primates and birds, can be mimicked in a simple model consisting of a system of interacting particles. The trick is that the system is guided by a constraint: It acts in a way that maximizes the amount of entropy (in this case, defined in terms of the different possible paths the particles could take) it generates within a given timespan.

Entropy maximization has long been thought to be a trait of nonequilibrium systems.[5] But the system in this model obeys a rule that lets it maximize entropy over a fixed time window that stretches into the future. In other words, it has foresight. In effect, the model looks at all the paths the particles could take and compels them to adopt the path that produces the greatest entropy. Crudely speaking, this tends to be the path that keeps open the largest number of options for how the particles might move subsequently.

You might say that the system of particles experiences a kind of urge to preserve freedom of future action, and that this urge guides its behavior at any moment. The researchers who developed the model—Alexander Wissner-Gross at Harvard University and Cameron Freer, a mathematician at the Massachusetts Institute of Technology—call this a "causal entropic force."[6] In computer simulations of configurations of disk-shaped particles moving around in particular settings, this force creates outcomes that are eerily suggestive of intelligence.

In one case, a large disk was able to "use" a small disk to extract a second small disk from a narrow tube—a process that looked like tool use. Freeing the disk increased the entropy of the system. In another example, two disks in separate compartments synchronized their behavior to pull a larger disk down so that they could interact with it, giving the appearance of social cooperation.

Of course, these simple interacting agents get the benefit of a glimpse into the future. Life, as a general rule, does not. So how relevant is this for biology? That's not clear, although Wissner-Gross said that he is now working to establish "a practical, biologically plausible, mechanism for causal entropic forces." In the meantime, he thinks that the approach could have practical spinoffs, offering a shortcut to artificial intelligence. "I predict that a faster way to achieve it will be to discover such behavior first and then work backward from the physical principles and constraints, rather than working forward from particular calculation or prediction techniques," he said. In other words, first find a system that does what you want it to do and then figure out how it does it.

Aging, too, has conventionally been seen as a trait dictated by evolution. Organisms have a lifespan that creates opportunities to reproduce, the story goes, without inhibiting the survival prospects of offspring by the parents sticking around too long and competing for resources. That seems surely to be part of the story, but Hildegard Meyer-Ortmanns, a physicist at Jacobs University in Bremen, Germany, thinks that ultimately aging is a physical process, not a biological one, governed by the thermodynamics of information.

It's certainly not simply a matter of things wearing out. "Most of the soft material we are made of is renewed before it has the chance to age," Meyer-Ortmanns said. But this renewal process isn't perfect. The thermodynamics of information copying dictates that there must be a trade-off between precision and energy.[7] An organism has a finite supply of energy, so errors necessarily accumulate over time. The organism then has to spend an increasingly large amount of energy to repair these errors. The renewal process eventually yields copies too flawed to function properly; death follows.

Empirical evidence seems to bear that out. It has been long known that cultured human cells seem able to replicate no more than 40 to 60 times (called the Hayflick limit) before they stop and become senescent.[8] And recent observations of human longevity have suggested that there may be some fundamental reason why humans can't survive much beyond age 100.[9]

There's a corollary to this apparent urge for energy-efficient, organized, predictive systems to appear in a fluctuating nonequilibrium environment. We ourselves are such a system, as are all our ancestors back to the first primitive cell. And nonequilibrium thermodynamics seems to be telling us that this is just what matter does under such circumstances. In other words, the appearance of life on a planet like the early Earth, imbued with energy sources such as sunlight and volcanic activity that keep things churning out of equilibrium, starts to seem not an extremely unlikely event, as many scientists have assumed, but virtually inevitable. In 2006, Eric Smith and the late Harold Morowitz at the Santa Fe Institute argued that the thermodynamics of nonequilibrium systems makes the emergence of organized, complex systems much more likely on a prebiotic Earth far from equilibrium than it would be if the raw chemical ingredients were just sitting in a "warm little pond" (as Charles Darwin put it) stewing gently.[10]

In the decade since that argument was first made, researchers have added detail and insight to the analysis. Those qualities that Ernst Mayr thought essential to biology—meaning and intention—may emerge as a natural consequence of statistics and thermodynamics. And those general properties may in turn lead naturally to something like life.

At the same time, astronomers have shown us just how many worlds there are—by some estimates stretching into the billions—orbiting other stars in our galaxy. Many are far from equilibrium, and at least a few are Earth-like. And the same rules are surely playing out there, too.

IN NEWLY CREATED LIFE-FORM, A MAJOR MYSTERY

Emily Singer

P eel away the layers of a house—the plastered walls, the slate roof, the hardwood floors—and you're left with a frame, the skeletal form that makes up the core of any structure. Can we do the same with life? Can scientists pare down the layers of complexity to reveal the essence of life, the foundation on which biology is built?

That's what Craig Venter and his collaborators have attempted to do in a study published in March 2016 in the journal *Science*.[1] Venter's team painstakingly whittled down the genome of *Mycoplasma mycoides,* a bacterium that lives in cattle, to reveal a bare-bones set of genetic instructions capable of making life. The result is a tiny organism named syn3.0 that contains just 473 genes. (By comparison, *E. coli* has about 4,000 to 5,000 genes, and humans have roughly 20,000.)

Yet within those 473 genes lies a gaping hole. Scientists have little idea what roughly a third of them do. Rather than illuminating the essential components of life, syn3.0 has revealed how much we have left to learn about the very basics of biology.

"To me, the most interesting thing is what it tells us about what we don't know," said Jack Szostak, a biochemist at Harvard University who was not involved in the study. "So many genes of unknown function seem to be essential."

"We were totally surprised and shocked," said Venter, a biologist who heads the J. Craig Venter Institute in La Jolla, California, and Rockville, Maryland, and is most famous for his role in mapping the human genome. The researchers had expected some number of unknown genes in the mix, perhaps totaling five to 10 percent of the genome. "But this is truly a stunning number," he said.

The seed for Venter's quest was planted in 1995, when his team deciphered the genome of *Mycoplasma genitalium,* a microbe that lives in the human urinary tract.[2] When Venter's researchers started work on this new project, they chose *M. genitalium*—the second complete bacterial genome to be sequenced—expressly for its diminutive genome size. With 517 genes and 580,000 DNA letters, it has one of the smallest known genomes in a self-replicating organism. (Some symbiotic microbes can survive with just 100-odd genes, but they rely on resources from their host to survive.)

M. genitalium's trim package of DNA raised the question: What is the smallest number of genes a cell could possess? "We wanted to know the basic gene components of life," Venter said. "It seemed like a great idea 20 years ago—we had no idea it would be a 20-year process to get here."

MINIMAL DESIGN

Venter and his collaborators originally set out to design a stripped-down genome based on what scientists knew about biology. They would start with genes involved in the most critical processes of the cell, such as copying and translating DNA, and build from there.

But before they could create this streamlined version of life, the researchers had to figure out how to design and build genomes from scratch. Rather than editing DNA in a living organism, as most researchers did, they wanted to exert greater control—to plan their genome on a computer and then synthesize the DNA in test tubes.

In 2008, Venter and his collaborator Hamilton Smith created the first synthetic bacterial genome by building a modified version of *M. genitalium*'s DNA.[3] Then in 2010 they made the first self-replicating synthetic organism, manufacturing a version of *M. mycoides'* genome and then transplanting it into a different *Mycoplasma* species.[4] The synthetic genome took over the cell, replacing the native operating system with a human-made version. The synthetic *M. mycoides* genome was mostly identical to the natural version, save for a few genetic watermarks—researchers added their names and a few famous quotes, including a slightly garbled version of Richard Feynman's assertion, "What I cannot create, I do not understand."

With the right tools finally in hand, the researchers designed a set of genetic blueprints for their minimal cell and then tried to build them. Yet "not one design worked," Venter said. He saw their repeated failures as a rebuke for their hubris. Does modern science have sufficient knowledge of basic biological principles to build a cell? "The answer was a resounding no," he said.

So the team took a different and more labor-intensive tack, replacing the design approach with trial and error. They disrupted *M. mycoides'* genes, determining which were essential for the bacteria to survive. They erased the extraneous genes to create syn3.0, which has a smaller genome than any independently replicating organism discovered on Earth to date.

What's left after trimming the genetic fat? The majority of the remaining genes are involved in one of three functions: producing RNA and proteins, preserving the fidelity of genetic information or creating the cell membrane. Genes for editing DNA were largely expendable.

But it is unclear what the remaining 149 genes do. Scientists can broadly classify 70 of them based on the genes' structure, but the researchers have little idea of what precise role the genes play in the cell. The function of 79 genes is a complete mystery. "We don't know what they provide or why they are essential for life—maybe they are doing something more subtle, something obviously not appreciated yet in biology," Venter said. "It's a very humbling set of experiments."

Venter's team is eager to figure out what the mystery genes do, but the challenge is multiplied by the fact that these genes don't resemble any other known genes. One way to investigate their function is to engineer versions of the cell in which each of these genes can be turned on and off. When they're off, "what's the first thing to get messed up?" Szostak said. "You can try to pin it to general class, like metabolism or DNA replication."

DWINDLING TO ZERO

Venter is careful to avoid calling syn3.0 a universal minimal cell. If he had done the same set of experiments with a different microbe, he points out, he would have ended up with a different set of genes.

In fact, there's no single set of genes that all living things need in order to exist. When scientists first began searching for such a thing about 20 years ago, they hoped that simply comparing the genome sequences from a bunch of different species would reveal an essential core shared by all species. But as the number of genome sequences blossomed, that essential core disappeared. In 2010, David Ussery, then a biologist at Oak Ridge National Laboratory in Tennessee, and his collaborators compared 1,000 genomes.[5] They found that not a single gene is shared across all of life. "There are different ways to have a core set of instructions," Szostak said.

Moreover, what's essential in biology depends largely on an organism's environment. For example, imagine a microbe that lives in the presence of a toxin, such as an antibiotic. A gene that can break down the toxin would be essential for a microbe in that environment. But remove the toxin, and that gene is no longer essential.

Venter's minimal cell is a product not just of its environment, but of the entirety of the history of life on Earth. Sometime in biology's 4-billion-year record, cells much simpler than this one must have existed. "We didn't go from nothing to a cell with 400 genes," Szostak said. He and others are trying to make more basic life-forms that are representative of these earlier stages of evolution.

Some scientists say that this type of bottom-up approach is necessary in order to truly understand life's essence. "If we are ever to understand even the simplest living organism, we have to be able to design and synthesize one from scratch," said Anthony Forster, a biologist at Uppsala University in Sweden. "We are still far from this goal."

BREAKTHROUGH DNA EDITOR BORN OF BACTERIA

Carl Zimmer

On a November evening in 2014, Jennifer Doudna put on a stylish black evening gown and headed to Hangar One, a building at NASA's Ames Research Center that was constructed in 1932 to house dirigibles. Under the looming arches of the hangar, Doudna mingled with celebrities like Benedict Cumberbatch, Cameron Diaz and Jon Hamm before receiving the 2015 Breakthrough Prize in life sciences, an award sponsored by Mark Zuckerberg and other tech billionaires. Doudna, a biochemist at the University of California, Berkeley, and her collaborator, Emmanuelle Charpentier, then of the Helmholtz Centre for Infection Research in Germany, each received $3 million for their invention of a potentially revolutionary tool for editing DNA known as CRISPR.[1]

Doudna was not a gray-haired emerita being celebrated for work she did back when dirigibles ruled the sky. It was only in 2012 that Doudna, Charpentier and their colleagues offered the first demonstration of CRISPR's potential. They crafted molecules that could enter a microbe and precisely snip its DNA at a location of the researchers' choosing. In January 2013, the scientists went one step further: They cut out a particular piece of DNA in human cells and replaced it with another one.

In the same month, separate teams of scientists at Harvard University and the Broad Institute reported similar success with the gene-editing tool. A scientific stampede commenced, and in just the past five years, researchers have performed hundreds of experiments on CRISPR. Their results hint that the technique may fundamentally change both medicine and agriculture.

Some scientists have repaired defective DNA in mice, for example, curing them of genetic disorders. Plant scientists have used CRISPR to edit genes in crops, raising hopes that they can engineer a better food supply. Some researchers are trying to rewrite the genomes of elephants, with the ultimate goal of re-creating a woolly mammoth. Writing in the journal *Reproductive Biology and Endocrinology*, Motoko Araki and Tetsuya Ishii of Hokkaido

University in Japan predicted that doctors will be able to use CRISPR to alter the genes of human embryos "in the immediate future."[2]

Thanks to the speed of CRISPR research, the accolades have come quickly. In 2014, *MIT Technology Review* called CRISPR "the biggest biotech discovery of the century." The Breakthrough Prize is just one of several prominent awards Doudna has won for her work on CRISPR; in May 2018, she, Charpentier, and Virginijus Šikšnys won the Kavli Prize for nanoscience.

Even the pharmaceutical industry, which is often slow to embrace new scientific advances, is rushing to get in on the act. New companies developing CRISPR-based medicine are opening their doors. In January 2015, the pharmaceutical giant Novartis announced that it would be using Doudna's CRISPR technology for its research into cancer treatments. More recently, the University of Pennsylvania confirmed plans for a clinical trial in which CRISPR would be used to edit the genes of human immune cells to make them attack tumors.

But amid all the black-tie galas and patent filings, it's easy to overlook the most important fact about CRISPR: Nobody actually invented it.

Doudna and other researchers did not pluck the molecules they use for gene editing from thin air. In fact, they stumbled across the CRISPR molecules in nature. Microbes have been using them to edit their own DNA for millions of years, and today they continue to do so all over the planet, from the bottom of the sea to the recesses of our own bodies.

We've barely begun to understand how CRISPR works in the natural world. Microbes use it as a sophisticated immune system, allowing them to learn to recognize their enemies. Now scientists are discovering that microbes use CRISPR for other jobs as well. The natural history of CRISPR poses many questions to scientists, for which they don't have very good answers yet. But it also holds great promise. Doudna and her colleagues harnessed one type of CRISPR, but scientists are finding a vast menagerie of different types. Tapping that diversity could lead to more effective gene editing technology, or open the way to applications no one has thought of yet.

"You can imagine that many labs—including our own—are busily looking at other variants and how they work," Doudna said. "So stay tuned."

A REPEAT MYSTERY

The scientists who discovered CRISPR had no way of knowing that they had discovered something so revolutionary. They didn't even understand what they had found. In 1987, Yoshizumi Ishino and colleagues at Osaka University in Japan published the sequence of a gene called *iap* belonging to the gut microbe *E. coli*. To better understand how the gene worked, the scientists also sequenced some of the DNA surrounding it. They hoped to find spots where proteins landed, turning *iap* on and off. But instead of a switch, the scientists found something incomprehensible.

Near the *iap* gene lay five identical segments of DNA. DNA is made up of building blocks called bases, and the five segments were each composed of the same 29 bases. These repeat sequences were separated from each other by 32-base blocks of DNA, called spacers. Unlike the repeat sequences, each of the spacers had a unique sequence.

This peculiar genetic sandwich didn't look like anything biologists had found before. When the Japanese researchers published their results, they could only shrug. "The biological significance of these sequences is not known," they wrote.

It was hard to know at the time if the sequences were unique to *E. coli*, because microbiologists only had crude techniques for deciphering DNA. But in the 1990s, technological advances allowed them to speed up their sequencing. By the end of the decade, microbiologists could scoop up seawater or soil and quickly sequence much of the DNA in the sample. This technique—called metagenomics—revealed those strange genetic sandwiches in a staggering number of species of microbes. They became so common that scientists needed a name to talk about them, even if they still didn't know what the sequences were for. In 2002, Ruud Jansen of Utrecht University in the Netherlands and colleagues dubbed these sandwiches "clustered regularly interspaced short palindromic repeats"—CRISPR for short.[3]

Jansen's team noticed something else about CRISPR sequences: They were always accompanied by a collection of genes nearby. They called these genes *Cas* genes, for CRISPR-associated genes. The genes encoded enzymes that could cut DNA, but no one could say why they did so, or why they always sat next to the CRISPR sequence.

Three years later, three teams of scientists independently noticed something odd about CRISPR spacers. They looked a lot like the DNA of viruses.

"And then the whole thing clicked," said Eugene Koonin.

At the time, Koonin, an evolutionary biologist at the National Center for Biotechnology Information in Bethesda, Maryland, had been puzzling over CRISPR and *Cas* genes for a few years. As soon as he learned of the discovery of bits of virus DNA in CRISPR spacers, he realized that microbes were using CRISPR as a weapon against viruses.

Koonin knew that microbes are not passive victims of virus attacks. They have several lines of defense. Koonin thought that CRISPR and *Cas* enzymes provide one more. In Koonin's hypothesis, bacteria use *Cas* enzymes to grab fragments of viral DNA. They then insert the virus fragments into their own CRISPR sequences. Later, when another virus comes along, the bacteria can use the CRISPR sequence as a cheat sheet to recognize the invader.

Scientists didn't know enough about the function of CRISPR and *Cas* enzymes for Koonin to make a detailed hypothesis. But his thinking was provocative enough for a microbiologist named Rodolphe Barrangou to test it. To Barrangou, Koonin's idea was not just fascinating, but potentially a huge deal for his employer at the time, the yogurt maker Danisco. Danisco depended on bacteria to convert milk into yogurt, and sometimes entire cultures would be lost to outbreaks of bacteria-killing viruses. Now Koonin was suggesting that bacteria could use CRISPR as a weapon against these enemies.

To test Koonin's hypothesis, Barrangou and his colleagues infected the milk-fermenting microbe *Streptococcus thermophilus* with two strains of viruses. The viruses killed many of the bacteria, but some survived. When those resistant bacteria multiplied, their descendants turned out to be resistant too. Some genetic change had occurred. Barrangou and his colleagues found that the bacteria had stuffed DNA fragments from the two viruses into their spacers. When the scientists chopped out the new spacers, the bacteria lost their resistance.

Barrangou, now an associate professor at North Carolina State University, said that this discovery led many manufacturers to select for customized CRISPR sequences in their cultures, so that the bacteria could withstand virus outbreaks. "If you've eaten yogurt or cheese, chances are you've eaten CRISPR-ized cells," he said.

CUT AND PASTE

As CRISPR started to give up its secrets, Doudna got curious. She had already made a name for herself as an expert on RNA, a single-stranded cousin to DNA. Originally, scientists had seen RNA's main job as a messenger. Cells would make a copy of a gene using RNA, and then use that messenger RNA as a template for building a protein. But Doudna and other scientists illuminated many other jobs that RNA can do, such as acting as sensors or controlling the activity of genes.

In 2007, Blake Wiedenheft joined Doudna's lab as a postdoctoral researcher, eager to study the structure of *Cas* enzymes to understand how they worked. Doudna agreed to the plan—not because she thought CRISPR had any practical value, but just because she thought the chemistry might be cool. "You're not trying to get to a particular goal, except understanding," she said.

As Wiedenheft, Doudna and their colleagues figured out the structure of *Cas* enzymes, they began to see how the molecules worked together as a system. When a virus invades a microbe, the host cell grabs a little of the virus's genetic material, cuts open its own DNA and inserts the piece of virus DNA into a spacer.

As the CRISPR region fills with virus DNA, it becomes a molecular most-wanted gallery, representing the enemies the microbe has encountered. The microbe can then use this viral DNA to turn *Cas* enzymes into precision-guided weapons. The microbe copies the genetic material in each spacer into an RNA molecule. *Cas* enzymes then take up one of the RNA molecules and cradle it. Together, the viral RNA and the *Cas* enzymes drift through the cell. If they encounter genetic material from a virus that matches the CRISPR RNA, the RNA latches on tightly. The *Cas* enzymes then chop the DNA in two, preventing the virus from replicating.

As CRISPR's biology emerged, it began to make other microbial defenses look downright primitive. Using CRISPR, microbes could, in effect, program their enzymes to seek out any short sequence of DNA and attack it exclusively.

"Once we understood it as a programmable DNA-cutting enzyme, there was an interesting transition," Doudna said. She and her colleagues realized there might be a very practical use for CRISPR. Doudna recalls thinking, "Oh my gosh, this could be a tool."

It wasn't the first time a scientist had borrowed a trick from microbes to build a tool. Some microbes defend themselves from invasion by using molecules known as restriction enzymes. The enzymes chop up any DNA that isn't protected by molecular shields. The microbes shield their own genes, and then attack the naked DNA of viruses and other parasites. In the 1970s, molecular biologists figured out how to use restriction enzymes to cut DNA, giving birth to the modern biotechnology industry.

In the decades that followed, genetic engineering improved tremendously, but it couldn't escape a fundamental shortcoming: Restriction enzymes did not evolve to make precise cuts—only to shred foreign DNA. As a result, scientists who used restriction enzymes for biotechnology had little control over where their enzymes cut open DNA.

The CRISPR-*Cas* system, Doudna and her colleagues realized, had already evolved to exert just that sort of control.

To create a DNA-cutting tool, Doudna and her colleagues picked out the CRISPR-*Cas* system from *Streptococcus pyogenes*, the bacteria that cause strep throat. It was a system they already understood fairly well, having worked out the function of its main enzyme, called *Cas*9. Doudna and her colleagues figured out how to supply *Cas*9 with an RNA molecule that matched a sequence of DNA they wanted to cut. The RNA molecule then guided *Cas9* along the DNA to the target site, and then the enzyme made its incision.

Using two *Cas9* enzymes, the scientists could make a pair of snips, chopping out any segment of DNA they wanted. They could then coax a cell to stitch a new gene into the open space. Doudna and her colleagues thus invented a biological version of find-and-replace—one that could work in virtually any species they chose to work on.

As important as these results were, microbiologists were also grappling with even more profound implications of CRISPR. It showed them that microbes had capabilities no one had imagined before.

Before the discovery of CRISPR, all the defenses that microbes were known to use against viruses were simple, one-size-fits-all strategies. Restriction enzymes, for example, will destroy any piece of unprotected DNA. Scientists refer to this style of defense as innate immunity. We have innate immunity, too, but on top of that, we also use an entirely different immune system to fight pathogens: one that learns about our enemies.

This so-called adaptive immune system is organized around a special set of immune cells that swallow up pathogens and then present fragments of them, called antigens, to other immune cells. If an immune cell binds tightly to an antigen, the cell multiplies. The process of division adds some random changes to the cell's antigen receptor genes. In a few cases, the changes alter the receptor in a way that lets it grab the antigen even more tightly. Immune cells with the improved receptor then multiply even more.

This cycle results in an army of immune cells with receptors that can bind quickly and tightly to a particular type of pathogen, making them into precise assassins. Other immune cells produce antibodies that can also grab onto the antigens and help kill the pathogen. It takes a few days for the adaptive immune system to learn to recognize the measles virus, for instance, and wipe it out. But once the infection is over, we can hold onto these immunological memories. A few immune cells tailored to measles stay with us for our lifetime, ready to attack again.

CRISPR, microbiologists realized, is also an adaptive immune system. It lets microbes learn the signatures of new viruses and remember them. And while we need a complex network of different cell types and signals to learn to recognize pathogens, a single-celled microbe has all the equipment necessary to learn the same lesson on its own.

But how did microbes develop these abilities? Ever since microbiologists began discovering CRISPR-*Cas* systems in different species, Koonin and his colleagues have been reconstructing the systems' evolution. CRISPR-*Cas* systems use a huge number of different enzymes, but all of them have one enzyme in common, called *Cas*1. The job of this universal enzyme is to grab incoming virus DNA and insert it in CRISPR spacers. Recently, Koonin and his colleagues discovered what may be the origin of *Cas*1 enzymes.

Along with their own genes, microbes carry stretches of DNA called mobile elements that act like parasites. The mobile elements contain genes for enzymes that exist solely to make new copies of their own DNA, cut open their host's genome and insert the new copy. Sometimes mobile elements can jump from one host to another, either by hitching a ride with a virus or by other means, and spread through their new host's genome.

Koonin and his colleagues discovered that one group of mobile elements, called casposons, makes enzymes that are pretty much identical to *Cas*1. In a paper in *Nature Reviews Genetics*, Koonin and Mart Krupovic of the Pasteur Institute in Paris argued that the CRISPR-*Cas* system got its start when mutations transformed casposons from enemies into friends.[4] Their DNA-cutting enzymes became domesticated, taking on a new function: to store captured virus DNA as part of an immune defense.

While CRISPR may have had a single origin, it has blossomed into a tremendous diversity of molecules. Koonin is convinced that viruses are responsible for this. Once they faced CRISPR's powerful, precise defense, the viruses evolved evasions. Their genes changed sequence so that CRISPR couldn't latch onto them easily. And the viruses also evolved molecules that could block the *Cas* enzymes. The microbes responded by evolving in their turn. They acquired new strategies for using CRISPR that the viruses couldn't fight. Over many thousands of years, in other words, evolution behaved like a natural laboratory, coming up with new recipes for altering DNA.

THE HIDDEN TRUTH

To Konstantin Severinov, who holds joint appointments at Rutgers University and the Skolkovo Institute of Science and Technology in Russia, these explanations for CRISPR may turn out to be true, but they barely begin to account for its full mystery. In fact, Severinov questions whether fighting viruses is the chief function of CRISPR. "The immune function may be a red herring," he said.

Severinov's doubts stem from his research on the spacers of *E. coli*. He and other researchers have amassed a database of tens of thousands of *E. coli* spacers, but only a handful of them match any virus known to infect *E. coli*. You can't blame this dearth on our ignorance of *E. coli* or its viruses, Severinov argues, because they've been the workhorses of molecular biology for a century. "That's kind of mind-boggling," he said.

It's possible that the spacers came from viruses, but viruses that disappeared thousands of years ago. The microbes kept holding onto the spacers even when they no longer had to face these enemies. Instead, they used CRISPR for other tasks. Severinov speculates that a CRISPR sequence might act as a kind of genetic bar code. Bacteria that shared the same bar code could recognize each other as relatives and cooperate, while fighting off unrelated populations of bacteria.

But Severinov wouldn't be surprised if CRISPR also carries out other jobs. Recent experiments have shown that some bacteria use CRISPR to silence their own genes, instead of seeking out the genes of enemies. By silencing their genes, the bacteria stop making molecules on their surface that are easily detected by our immune system. Without this CRISPR cloaking system, the bacteria would blow their cover and get killed.

"This is a fairly versatile system that can be used for different things," Severinov said, and the balance of all those things may differ from system to system and from species to species.

If scientists can get a better understanding of how CRISPR works in nature, they may gather more of the raw ingredients for technological innovations. To create a new way to edit DNA, Doudna and her colleagues exploited the CRISPR-*Cas* system from a single species of bacteria, *Streptococcus pyogenes*. There's no reason to assume that it's the best system for that application. At Editas, a company based in Cambridge, Massachusetts, scientists have been investigating the *Cas*9 enzyme made by another species of bacteria, *Staphylococcus aureus*. In January 2015, Editas scientists reported that it's about as efficient at cutting DNA as *Cas*9 from *Streptococcus pyogenes*. But it also has some potential advantages, including its small size, which may make it easier to deliver into cells.

To Koonin, these discoveries are just baby steps into the ocean of CRISPR diversity. Scientists are now working out the structure of distantly related versions of *Cas*9 that seem to behave very differently from the ones we're now familiar with. "Who knows whether this thing could become even a better tool?" Koonin said.

And as scientists discover more tasks that CRISPR accomplishes in nature, they may be able to mimic those functions, too. Doudna is curious about using CRISPR as a diagnostic tool, searching cells for cancerous mutations, for example. "It's seek and detect, not seek and destroy," she said. But having been surprised by CRISPR before, Doudna expects the biggest benefits from these molecules to surprise us yet again. "It makes you wonder what else is out there," she said.

NEW LETTERS ADDED TO THE GENETIC ALPHABET

Emily Singer

D NA stores our genetic code in an elegant double helix. But some argue that this elegance is overrated. "DNA as a molecule has many things wrong with it," said Steven Benner, an organic chemist at the Foundation for Applied Molecular Evolution in Florida.

Nearly 30 years ago, Benner sketched out better versions of both DNA and its chemical cousin RNA, adding new letters and other additions that would expand their repertoire of chemical feats. He wondered why these improvements haven't occurred in living creatures. Nature has written the entire language of life using just four chemical letters: G, C, A and T. Did our genetic code settle on these four nucleotides for a reason? Or was this system one of many possibilities, selected by simple chance? Perhaps expanding the code could make it better.

Benner's early attempts at synthesizing new chemical letters failed. But with each false start, his team learned more about what makes a good nucleotide and gained a better understanding of the precise molecular details that make DNA and RNA work. The researchers' efforts progressed slowly, as they had to design new tools to manipulate the extended alphabet they were building. "We have had to re-create, for our artificially designed DNA, all of the molecular biology that evolution took 4 billion years to create for natural DNA," Benner said.

Now, after decades of work, Benner's team has synthesized artificially enhanced DNA that functions much like ordinary DNA, if not better. In two papers published in the *Journal of the American Chemical Society* in June 2015, the researchers have shown that two synthetic nucleotides called P and Z fit seamlessly into DNA's helical structure, maintaining the natural shape of DNA.[1] Moreover, DNA sequences incorporating these letters can evolve just like traditional DNA, a first for an expanded genetic alphabet.[2]

The new nucleotides even outperform their natural counterparts. When challenged to evolve a segment that selectively binds to cancer cells, DNA sequences using P and Z did better than those without.

"When you compare the four-nucleotide and six-nucleotide alphabet, the six-nucleotide version seems to have won out," said Andrew Ellington, a biochemist at the University of Texas, Austin, who was not involved in the study.

Benner has lofty goals for his synthetic molecules. He wants to create an alternative genetic system in which proteins—intricately folded molecules that perform essential biological functions—are unnecessary. Perhaps, Benner proposes, instead of our standard three-component system of DNA, RNA and proteins, life on other planets evolved with just two.

BETTER BLUEPRINTS FOR LIFE

The primary job of DNA is to store information. Its sequence of letters contains the blueprints for building proteins. Our current four-letter alphabet encodes 20 amino acids, which are strung together to create millions of different proteins. But a six-letter alphabet could encode as many as 216 possible amino acids and many, many more possible proteins.

Why nature stuck with four letters is one of biology's fundamental questions. Computers, after all, use a binary system with just two "letters"—0s and 1s. Yet two letters probably aren't enough to create the array of biological molecules that make up life. "If you have a two-letter code, you limit the number of combinations you get," said Ramanarayanan Krishnamurthy, a chemist at the Scripps Research Institute in La Jolla, California.

On the other hand, additional letters could make the system more error prone. DNA bases come in pairs—G pairs with C and A pairs with T. It's this pairing that endows DNA with the ability to pass along genetic information. With a larger alphabet, each letter has a greater chance of pairing with the wrong partner, and new copies of DNA might harbor more mistakes. "If you go past four, it becomes too unwieldy," Krishnamurthy said.

But perhaps the advantages of a larger alphabet can outweigh the potential drawbacks. Six-letter DNA could densely pack in genetic information. And perhaps six-letter RNA could take over some of the jobs now handled by proteins, which perform most of the work in the cell.

Proteins have a much more flexible structure than DNA and RNA and are capable of folding into an array of complex shapes. A properly folded protein can act as a molecular lock, opening a chamber only for the right key. Or it can act as a catalyst, capturing and bringing together different molecules for chemical reactions.

ATCG   ATCGPZ

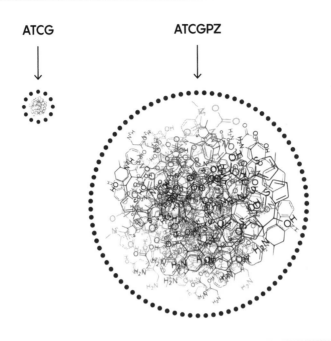

FIGURE 4.1

Expanding the genetic alphabet dramatically expands the number of possible amino acids and proteins that cells can build, at least in theory. The existing four-letter alphabet produces 20 amino acids (small circle) while a six-letter alphabet could produce 216 possible amino acids (large circle).

Adding new letters to RNA could give it some of these abilities. "Six letters can potentially fold into more, different structures than four letters," Ellington said.

Back when Benner was sketching out ideas for alternative DNA and RNA, it was this potential that he had in mind. According to the most widely held theory of life's origins, RNA once performed both the information-storage job of DNA and the catalytic job of proteins. Benner realized that there are many ways to make RNA a better catalyst.

"With just these little insights, I was able to write down the structures that are in my notebook as alternatives that would make DNA and RNA better," Benner said. "So the question is: Why did life not make these alternatives? One way to find out was to make them ourselves, in the laboratory, and see how they work."

It's one thing to design new codes on paper, and quite another to make them work in real biological systems. Other researchers have created their own additions to the genetic code, in one case even incorporating new letters into living bacteria.[3] But these other bases fit together a bit differently from natural ones, stacking on top of each other rather than linking side by side. This can distort the shape of DNA, particularly when a number of these bases cluster together. Benner's P-Z pair, however, is designed to mimic natural bases.

One of the papers by Benner's team shows that Z and P are yoked together by the same chemical bond that ties A to T and C to G.[4] (This bond is known as Watson-Crick pairing, after the scientists who discovered DNA's structure.) Millie Georgiadis, a chemist at Indiana University-Purdue University Indianapolis, along with Benner and other collaborators, showed that DNA strands that incorporate Z and P retain their proper helical shape if the new letters are strung together or interspersed with natural letters.

"This is very impressive work," said Jack Szostak, a Harvard chemist who studies the origin of life but was not involved in the study. "Finding a novel base pair that does not grossly disrupt the double-helical structure of DNA has been quite difficult."

The team's second paper demonstrates how well the expanded alphabet works.[5] Researchers started with a random library of DNA strands constructed from the expanded alphabet and then selected the strands that were able to bind to liver cancer cells but not to other cells. Of the 12 successful binders, the best had Zs and Ps in their sequences, while the weakest did not.

"More functionality in the nucleobases has led to greater functionality in nucleic acids themselves," Ellington said. In other words, the new additions appear to improve the alphabet, at least under these conditions.

But additional experiments are needed to determine how broadly that's true. "I think it will take more work, and more direct comparisons, to be sure that a six-letter version generally results in 'better' aptamers [short DNA strands] than four-letter DNA," Szostak said. For example, it's unclear whether the six-letter alphabet triumphed because it provided more sequence options or because one of the new letters is simply better at binding, Szostak said.

Benner wants to expand his genetic alphabet even further, which could enhance its functional repertoire. He's working on creating a 10- or 12-letter system and plans to move the new alphabet into living cells. Benner's and others' synthetic molecules have already proved useful in medical and biotech applications, such as diagnostic tests for HIV and other diseases. Indeed, Benner's work helped to found the burgeoning field of synthetic biology, which seeks to build new life, in addition to forming useful tools from molecular parts.

WHY LIFE'S CODE IS LIMITED

Benner's work and that of other researchers suggests that a larger alphabet has the capacity to enhance DNA's function. So why didn't nature expand its alphabet in the 4 billion years it has had to work on it? It could be because a larger repertoire has potential disadvantages. Some of the structures made possible by a larger alphabet might be of poor quality, with a greater risk of misfolding, Ellington said.

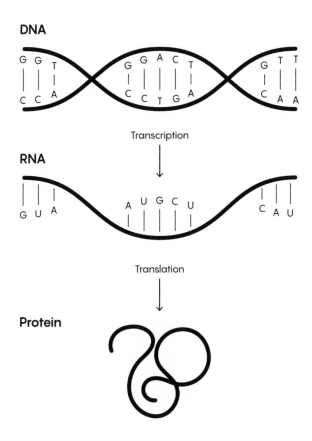

FIGURE 4.2

The genetic code—made up of the four letters, A, T, G and C—stores the blueprint for proteins. DNA is first transcribed into RNA and then translated into proteins, which fold into specific shapes.

Nature was also effectively locked into the system at hand when life began. "Once [nature] has made a decision about which molecular structures to place at the core of its molecular biology, it has relatively little opportunity to change those decisions," Benner said. "By constructing unnatural systems, we are learning not only about the constraints at the time that life first emerged, but also about constraints that prevent life from searching broadly within the imagination of chemistry."

Benner aims to make a thorough search of that chemical space, using his discoveries to make new and improved versions of both DNA and RNA. He wants to make DNA better at storing information and RNA better at catalyzing reactions. He hasn't shown directly that the P-Z base pairs do that. But both bases have the potential to help RNA fold into more complex structures, which in turn could make proteins better catalysts. P has a place to add a "functional group," a molecular structure that helps folding and is typically found in proteins. And Z has a nitro group, which could aid in molecular binding.

In modern cells, RNA acts as an intermediary between DNA and proteins. But Benner ultimately hopes to show that the three-biopolymer system—DNA, RNA and proteins—that exists throughout life on Earth isn't essential. With better-engineered DNA and RNA, he says, perhaps proteins are unnecessary.

Indeed, the three-biopolymer system may have drawbacks, since information flows only one way, from DNA to RNA to proteins. If a DNA mutation produces a more efficient protein, that mutation will spread slowly, as organisms without it eventually die off.

What if the more efficient protein could spread some other way, by directly creating new DNA? DNA and RNA can transmit information in both directions. So a helpful RNA mutation could theoretically be transformed into beneficial DNA. Adaptations could thus lead directly to changes in the genetic code.

Benner predicts that a two-biopolymer system would evolve faster than our own three-biopolymer system. If so, this could have implications for life on distant planets. "If we find life elsewhere," he said, "it would likely have the two-biopolymer system."

THE SURPRISING ORIGINS OF LIFE'S COMPLEXITY

Carl Zimmer

C harles Darwin was not yet 30 when he got the basic idea for the theory of evolution. But it wasn't until he turned 50 that he presented his argument to the world. He spent those two decades methodically compiling evidence for his theory and coming up with responses to every skeptical counterargument he could think of. And the counterargument he anticipated most of all was that the gradual evolutionary process he envisioned could not produce certain complex structures.

Consider the human eye. It is made up of many parts—a retina, a lens, muscles, jelly and so on—all of which must interact for sight to occur. Damage one part—detach the retina, for instance—and blindness can follow. In fact, the eye functions only if the parts are of the right size and shape to work with one another. If Darwin was right, then the complex eye had evolved from simple precursors. In *On the Origin of Species,* Darwin wrote that this idea "seems, I freely confess, absurd in the highest possible degree."

But Darwin could nonetheless see a path to the evolution of complexity. In each generation, individuals varied in their traits. Some variations increased their survival and allowed them to have more offspring. Over generations those advantageous variations would become more common—would, in a word, be "selected." As new variations emerged and spread, they could gradually tinker with anatomy, producing complex structures.

The human eye, Darwin argued, could have evolved from a simple light-catching patch of tissue of the kind that animals such as flatworms grow today. Natural selection could have turned the patch into a cup that could detect the direction of the light. Then, some added feature would work with the cup to further improve vision, better adapting an organism to its surroundings, and so this intermediate precursor of an eye would be passed down to future generations. And, step by step, natural selection could drive this transformation to increased complexity because each intermediate form would provide an advantage over what came before.

Darwin's musings on the origin of complexity have found support in modern biology. Today biologists can probe the eye and other organs in detail at the molecular level, where they find immensely complex proteins joining together to make structures that bear a striking resemblance to portals, conveyor belts and motors. Such intricate systems of proteins can evolve from simpler ones, with natural selection favoring the intermediates along the way.

But some scientists and philosophers have suggested that complexity can arise through other routes. Some argue that life has a built-in tendency to become more complex over time. Others maintain that as random mutations arise, complexity emerges as a side effect, even without natural selection to help it along. Complexity, they say, is not purely the result of millions of years of fine-tuning through natural selection—the process that Richard Dawkins famously dubbed "the blind watchmaker." To some extent, it just happens.

A SUM OF VARIED PARTS

Biologists and philosophers have pondered the evolution of complexity for decades, but according to Daniel W. McShea, a paleobiologist at Duke University, they have been hobbled by vague definitions. "It's not just that they don't know how to put a number on it. They don't know what they mean by the word," McShea said.

McShea has been contemplating this question for years, working closely with Robert N. Brandon, also at Duke. McShea and Brandon suggest that we look not only at the sheer number of parts making up living things but at the types of parts. Our bodies are made of 10 trillion cells. If they were all of retina, for example, has about 60 different kinds of neurons, each with a distinct task. By this measure, we can say that we humans are, indeed, more complex than an animal such as a sponge, which has perhaps only six cell types.

One advantage of this definition is that you can measure complexity in many ways. Our skeletons have different types of bones, for example, each with a distinctive shape. Even the spine is made up of different types of parts, from the vertebrae in the neck that hold up our head to the ones that support our rib cage.

In their 2010 book *Biology's First Law*, McShea and Brandon outlined a way that complexity defined in this way could arise. They argued that a bunch of parts that start out more or less the same should differentiate over time. Whenever organisms reproduce, one or more of their genes may mutate. And sometimes these mutations give rise to more types of parts. Once an organism has more parts, those units have an opportunity to become different. After a gene is accidentally copied, the duplicate may pick up mutations that the original does not share. Thus, if you start with a set of identical parts, according to McShea and Brandon, they will tend to become increasingly different from one another. In other words, the organism's complexity will increase.

As complexity arises, it may help an organism survive better or have more offspring. If so, it will be favored by natural selection and spread through the population. Mammals, for example, smell by binding odor molecules to receptors on nerve endings in their nose. These receptor genes have repeatedly duplicated over millions of years. The new copies mutate, allowing mammals to smell a wider range of aromas. Animals that rely heavily on their nose, such as mice and dogs, have more than 1,000 of these receptor genes. On the other hand, complexity can be a burden. Mutations can change the shape of a neck vertebra, for instance, making it hard for the head to turn. Natural selection will keep these mutations from spreading through populations. That is, organisms born with those traits will tend to die before reproducing, thus taking the deleterious traits out of circulation when they go. In these cases, natural selection works against complexity.

Unlike standard evolutionary theory, McShea and Brandon see complexity increasing even in the absence of natural selection. This statement is, they maintain, a fundamental law of biology—perhaps its only one. They have dubbed it the zero-force evolutionary law.

THE FRUIT-FLY TEST

McShea and Leonore Fleming, then a graduate student at Duke, put the zero-force evolutionary law to the test. The subjects were *Drosophila* flies. For more than a century scientists have reared stocks of the flies to use in experiments. In their laboratory homes, the flies have led a pampered life, provided with a constant supply of food and a steady, warm climate. Their wild relatives, meanwhile, have to contend with starvation, predators, cold and heat. Natural selection is strong among the wild flies, eliminating mutations that make flies unable to cope with their many challenges. In the sheltered environment of the labs, in contrast, natural selection is feeble.

The zero-force evolutionary law makes a clear prediction: over the past century the lab flies should have been less subject to the elimination of disadvantageous mutations and thus should have become more complex than the wild ones.

Fleming and McShea examined the scientific literature for 916 laboratory lines of flies. They made many different measures of complexity in each population. In the journal *Evolution & Development,* they reported that the lab flies were indeed more complex than wild ones.[1]

Although some biologists have endorsed the zero-force evolutionary law, Douglas Erwin, a leading paleontologist at the Smithsonian National Museum of Natural History, thinks it has some serious flaws. "One of its basic assumptions fails," he argues. According to the law, complexity may increase in the absence of selection. But that would be true only if organisms could actually exist beyond the influence of selection. In the real world, even when they are pampered by the most doting of scientists, Erwin contends, selection still exerts a force. For an animal such as a fly to develop properly, hundreds of genes have to interact in an elaborate choreography, turning one cell into many, giving rise to different organs and so on. Mutations may disrupt that choreography, preventing the flies from becoming viable adults.

An organism can exist without external selection—without the environment determining who wins and loses in the evolutionary race—but it will still be subject to internal selection, which takes place within organisms. In their study, McShea and Fleming did not provide evidence for the zero-force evolutionary law, according to Erwin, "because they only consider adult variants." The researchers did not look at the mutants that died from developmental disorders before reaching maturity, despite being cared for by scientists.

Some of the insects had irregular legs. Others acquired complicated patterns of colors on their wings. The segments of their antennae took on different shapes. Freed from natural selection, flies have reveled in complexity.

Another objection Erwin and other critics have raised is that McShea and Brandon's version of complexity does not jibe with how most people define the term. After all, an eye does not just have many different parts. Those parts also carry out a task together, and each one has a particular job to do. But McShea and Brandon argue that the kind of complexity that they are examining could lead to complexity of other sorts. "The kind of complexity that we're seeing in this *Drosophila* population is the foundation for really interesting stuff that selection could get hold of" to build complex structures that function to aid survival, McShea said.

MOLECULAR COMPLEXITY

As a paleobiologist, McShea is accustomed to thinking about the kind of complexity he can see in fossils—bones fitting together into a skeleton, for example. But in recent years a number of molecular biologists have independently begun to think much as he does about how complexity emerges.

In the 1990s a group of Canadian biologists started to ponder the fact that mutations often have no effect on an organism at all. These mutations are, in the jargon of evolutionary biology, neutral. The scientists, including Michael Gray of Dalhousie University in Halifax, proposed that the mutations could give rise to complex structures without going through a series of intermediates that are each selected for their help in adapting an organism to its environment. They dubbed this process "constructive neutral evolution."[2]

Gray has been encouraged by more recent studies that provide compelling evidence for constructive neutral evolution. One of the leaders in this research is Joe Thornton of the University of Oregon. He and his colleagues have found what appears to be an example in the cells of fungi. In fungi, such as a portobello mushroom, cells have to move atoms from one place to another to stay alive. One of the ways they do so is with molecular pumps called vacuolar ATPase complexes. A spinning ring of proteins shuttles atoms from one side of a membrane in the fungus to another. This ring is clearly a complex structure. It contains six protein molecules. Four of the molecules consist of the protein known as Vma3. The fifth is Vma11 and the sixth Vma16. All three types of protein are essential for the ring to spin.

To find out how this complex structure evolved, Thornton and his colleagues compared the proteins with related versions in other organisms, such as animals. (Fungi and animals share a common ancestor that lived around a billion years ago.)

In animals, the vacuolar ATPase complexes also have spinning rings made of six proteins. But those rings are different in one crucial way: instead of having three types of proteins in their rings, they have only two. Each animal ring is made up of five copies of Vma3 and one of Vma16. They have no Vma11. By McShea and Brandon's definition of complexity, fungi are more complex than animals—at least when it comes to their vacuolar ATPase complexes.

The scientists looked closely at the genes encoding the ring proteins. Vma11, the ring protein unique to fungi, turns out to be a close relative of the Vma3 in both animals and fungi. The genes for Vma3 and Vma11 must therefore share a common ancestry. Thornton and his colleagues concluded that early in the evolution of fungi, an ancestral gene for ring proteins was accidentally duplicated. Those two copies then evolved into Vma3 and Vma11.

By comparing the differences in the genes for Vma3 and Vma11, Thornton and his colleagues reconstructed the ancestral gene from which they both evolved. They then used that DNA sequence to create a corresponding protein—in effect, resurrecting an 800-million-year-old protein. The scientists called this protein Anc.3–11—short for *anc*estor of Vma*3* and Vma*11*. They wondered how the protein ring functioned with this ancestral protein. To find out, they inserted the gene for Anc.3–11 into the DNA of yeast. They also shut down its descendant genes, Vma3 and Vma11. Normally, shutting down the genes for the Vma3 and Vma11 proteins would be fatal because the yeast could no longer make their rings. But Thornton and his co-workers found that the yeast could survive with Anc.3–11 instead. It combined Anc.3–11 with Vma16 to make fully functional rings.

Experiments such as this one allowed the scientists to formulate a hypothesis for how the fungal ring became more complex. Fungi started out with rings made from only two proteins—the same ones found in animals like us. The proteins were versatile, able to bind to themselves or to their partners, joining up to proteins either on their right or on their left. Later the gene for Anc.3–11 duplicated into Vma3 and Vma11. These new proteins kept doing what the old ones had done: they assembled into rings for pumps. But over millions of generations of fungi, they began to mutate. Some of those mutations took away some of their versatility. Vma11, for example, lost the ability to bind to Vma3 on its clockwise side. Vma3 lost the ability to bind to Vma16 on its clockwise side. These mutations did not kill the yeast, because the proteins could still link together into a ring. They were neutral mutations, in other words. But now the ring had to be more complex because it could form successfully only if all three proteins were present and only if they arranged themselves in one pattern.

Thornton and his colleagues have uncovered precisely the kind of evolutionary episode predicted by the zero-force evolutionary law. Over time, life produced more parts—that is, more ring proteins. And then those extra parts began to diverge from one another. The fungi ended up with a more complex structure than their ancestors had. But it did not happen the way Darwin had imagined, with natural selection favoring a series of intermediate forms. Instead the fungal ring degenerated its way into complexity.

FIXING MISTAKES

Gray has found another example of constructive neutral evolution in the way many species edit their genes.[3] When cells need to make a given protein, they transcribe the DNA of its gene into RNA, the single-stranded counterpart of DNA, and then use special enzymes to replace certain RNA building blocks (called nucleotides) with other ones. RNA editing is essential to many species, including us—the unedited RNA molecules produce proteins that do not work. But there is also something decidedly odd about it. Why don't we just have genes with the correct original sequence, making RNA editing unnecessary?

The scenario that Gray proposes for the evolution of RNA editing goes like this: an enzyme mutates so that it can latch onto RNA and change certain nucleotides. This enzyme does not harm the cell, nor does it help it—at least not at first. Doing no harm, it persists. Later a harmful mutation occurs in a gene. Fortunately, the cell already has the RNA-binding enzyme, which can compensate for this mutation by editing the RNA. It shields the cell from the harm of the mutation, allowing the mutation to get passed down to the next generation and spread throughout the population. The evolution of this RNA-editing enzyme and the mutation it fixed was not driven by natural selection, Gray argues. Instead this extra layer of complexity evolved on its own—"neutrally." Then, once it became widespread, there was no way to get rid of it.

David Speijer, a biochemist at the University of Amsterdam, thinks that Gray and his colleagues have done biology a service with the idea of constructive neutral evolution, especially by challenging the notion that all complexity must be adaptive. But Speijer worries they may be pushing their argument too hard in some cases. On one hand, he thinks that the fungus pumps are a good example of constructive neutral evolution. "Everybody in their right mind would totally agree with it," he said. In other cases, such as RNA editing, scientists should not, in his view, dismiss the possibility that natural selection was at work, even if the complexity seems useless.

Gray, McShea and Brandon acknowledge the important role of natural selection in the rise of the complexity that surrounds us, from the biochemistry that builds a feather to the photosynthetic factories inside the leaves of trees. Yet they hope their research will coax other biologists to think beyond natural selection and to see the possibility that random mutation can fuel the evolution of complexity on its own. "We don't dismiss adaptation at all as part of that," Gray said. "We just don't think it explains everything."

ANCIENT SURVIVORS COULD REDEFINE SEX

Emily Singer

I f all the animals on Earth could offer a single lesson for long-term survival, it might be this: Sex works. Out of the estimated 8 million animal species, all but a smattering are known to reproduce sexually, and those that don't are babes in evolutionary terms, newly evolved animals that recently lost the ability to mate. "Sex must be important—if you lose it, you go extinct," said David Mark Welch, a biologist at the Marine Biological Laboratory in Woods Hole, Massachusetts.

Yet even though sex is the overwhelmingly dominant method of animal reproduction, scientists aren't sure why that is. Mark Welch estimates that researchers have developed around 50 to 60 hypotheses to explain the primacy of sex in the animal kingdom. Some of these theories have been biological battlegrounds for more than a century.

Studying the exception could help scientists understand the rule. And the exception in this case is a class of creatures called bdelloid rotifers (the "b" is silent), microscopic swimmers that split off from their sexual ancestors 40 million to 100 million years ago.

These bizarre animals are chaste survivors in a carnal world. They can withstand more radiation than any other animal tested to date. They can inhabit any surface that gets wet, from damp tree lichens to evaporated birdbaths. And without water, they hunker down in a state of total desiccation and then snap back to life with just a drop of liquid.

An analysis of the bdelloid genome[1] has begun to reveal how asexual mechanisms can mimic the DNA swapping characteristic of sex, perhaps even surpassing it in effectiveness. The new work has shown bdelloids to be so good at generating genetic diversity that some researchers now question the very definition of sex, with some arguing for a more expansive one that doesn't require the orchestrated swapping of genetic material. Others think that even if the traditional definition of sex remains intact, the unique genetic strategies of the bdelloid rotifer will illuminate the mechanisms that make sex such a successful evolutionary strategy. "If we can figure out the problem that bdelloids solved," said Mark Welch, who has been studying rotifers since the 1980s, "we can figure out why sex is important."

WHY SEX IS POPULAR

Reduced to its most basic form, sex is about the exchange of DNA. At the heart of this transaction is a process called meiosis, where chromosomes inherited from each parent pair up and swap pieces. The chromosomes are then divided among daughter cells. The result is a set of cells whose genome is different from that of either parent.

The benefits of this exchange seem obvious. The genetic shuffle creates a diverse population, and a diverse population should be better able to cope with a changing environment. This basic idea was first proposed by the German biologist August Weismann more than a century ago.

But sex also has substantial drawbacks, presenting something of a puzzle to evolutionary biologists. A sexual organism passes on only half of its genes, which significantly reduces its genetic legacy. And because sex shuffles the genome, it breaks up genetic combinations that work well. In addition, an animal that wants to mate must spend time and energy searching for a mate. Once that match is found, the act of sex carries the risk of sexually transmitted diseases, a very real danger in the natural world.

Given the drawbacks of sexual reproduction, we might expect the animal kingdom to be filled with both sexual and asexual creatures. But this is not the case; sex overwhelmingly predominates. "After hundreds of years, we still don't know what's so important about it," Mark Welch said. "One of the big quandaries is the contrast between the apparent short-term advantages of asexuality versus the apparent long-term advantages of sex—how do you even get the chance to reap the long-term benefits?"

Out of all the hypotheses biologists have developed, Weismann's basic premise—that sex gives animals the variation needed to deal with a changing environment—is still a top contender. In the century since he proposed it, theoretical biologists have devised specific mechanisms that would explain why it works. For example, sex might unite two important adaptations. One group of animals might develop a tolerance to high temperatures, for example, another to a specific toxin. Without sex, those two capabilities would be unlikely to come together in one species.

According to a hypothesis known as the Red Queen, which is sometimes considered a variant of Weismann's proposal, sex might help animals in their eternal arms race against pathogens. The genetic shuffling in sexual reproduction would help them quickly evolve defenses against rapidly morphing enemies. (The name of the hypothesis derives from a passage in *Through the Looking-Glass,* by Lewis Carroll, in which the Red Queen tells Alice to run as fast as she can in order to stay in the same place.)

Another theory, called Muller's ratchet, first put forth by the geneticist Hermann Muller in the 1960s, suggests that sexual reproduction helps rid the genome of harmful mistakes. In asexual organisms, new mutations occur in each generation and are passed on to the next, eventually driving the species to extinction. (It's called a ratchet because, in theory, once the genome develops an error, it's stuck—there is no plausible way back.) The genetic shuffling that occurs during sex could act like a dust cloth to wipe away the offending mutations.

Scientists have accumulated evidence in support of each of these hypotheses. Yet researchers find it difficult to test any of them directly. Bdelloids offer a complementary approach. "Understanding how they cope without sex will help us understand why sex is important," said Diego Fontaneto, a biologist at the Institute of Ecosystem Study in Italy.

JUMBLED CHROMOSOMES

Bdelloids have been squirming beneath scientists' microscopes since 1696. In all that time, no one has spotted a male. (Sexual rotifer varieties have clearly distinguishable males, with a penis-like organ and sperm.) No one thought much about this curious void for nearly 200 years, until biologists first began to study asexual reproduction in animals, Mark Welch said.

Such a long absence of males is telling, but it's not definitive proof of asexuality. Other organisms that were once thought to go without intercourse were later found to mate under rare circumstances, often triggered by stress. "There have been many putative asexuals, but when people looked more closely, they did find some kind of secret sex going on," Mark Welch said.

Beginning in the late 1980s, Matthew Meselson, a renowned biologist at Harvard University, argued that perhaps the bdelloid genome could be used to test the organisms' asexuality. Most animals have chromosomes made of two nearly identical copies of each gene, a consequence of the pairing and mixing that goes on during meiosis. In asexual animals this mixing wouldn't happen, and the two copies should remain stubbornly distinct.

Just as the Human Genome Project was wrapping up in 2000, Meselson and Mark Welch, who had been Meselson's graduate student, published the first results of their exploration of the bdelloid genome. They found that bdelloids often had two very different copies of their genes.[2]

But the bdelloid genome would soon reveal even more interesting secrets. The animals often had not just two copies of a gene, as humans do, but four copies. Scientists began to suspect that at some point in bdelloids' evolutionary history, the entire genome replicated, leaving the creatures with an extra set of chromosomes.[3]

What were these chromosomes doing? To investigate, researchers had to sequence the entire genome. (Previously they had examined single genes or bits of chromosomes.) In 2009, a team that included Mark Welch and Karine Van Doninck, a biologist at the University of Namur in Belgium, received a grant to undertake the work. What they found was more intriguing than they anticipated.

The bdelloid genome is composed of more than just bdelloid genes. It is a Frankensteinian collage of foreign DNA.[4] Nearly 10 percent of the bdelloid's genome comes from outside the animal kingdom entirely, with fungi, plants and bacteria all contributing. This percentage is much higher than for other animals. In this regard, bdelloids more closely resemble bacteria, which frequently incorporate alien DNA into their genomes, a process known as horizontal gene transfer.

What's more, the bdelloid chromosomes are a jumble; bits and pieces of them have been moved around like a mismatched puzzle. "The highly rearranged chromosome was new and unexpected," said John Logsdon, an evolutionary biologist at the University of Iowa who was not involved in the project. "It's very unusual."

Nature sometimes jumbles a chromosome, but major rearrangements in sexually reproducing organisms render the unlucky individual sterile: If the maternally inherited chromosome is structured A-B-C, it can't pair with a paternally inherited chromosome that's ordered A-C-B. (Some hybrid species, such as mules, are sterile for a similar reason. The chromosomes from the horse mother and donkey father are mismatched.)

The full genome sequence provided the most direct evidence yet that bdelloid rotifers are asexual: No organism with such a mismatched set of chromosomes could possibly go through traditional meiosis. "Over millions of years, the genome has undergone so many rearrangements it's no longer possible for the chromosomes to pair," Mark Welch said.

These two striking properties—incorporating large amounts of alien DNA into their genomes and rearranging their own DNA—might help bdelloids surmount the problem of genetic diversity that plagues asexual animals. "There are quite a few ways in which asexual organisms can apparently overcome some of the disadvantages" of not having sex, said Bill Birky, an evolutionary geneticist at the University of Arizona, who was not involved in the sequencing project. Bdelloids' ability to take up foreign DNA can potentially give them new powers, allowing them to break down a toxin, for example.[5]

Copying and replacing pieces of their own chromosomes can sometimes boost the effect of beneficial mutations and remove harmful ones, defying Muller's ratchet.

Indeed, bdelloids appear to have adopted an evolutionary strategy similar to that of bacteria, a highly successful class of organisms that also lack conventional sex. "Researchers working on the evolutionary significance of sex often tended to overlook the fact that bacteria have been doing very well without sex for millions of years," said Jean-François Flot, a biologist now at the Free University of Brussels who participated in the bdelloid genome project.

Moreover, bdelloids' extra pair of chromosomes might generate additional genetic diversity. The redundant pair of chromosomes provides a new reservoir of genetic material that is free to evolve and take on new functions, which might help the bdelloids cope with changing environments in the future, Fontaneto said.

Yet not everyone is convinced that bdelloids are entirely asexual. "To me, the evidence is not completely slam-dunk in terms of demonstrating asexuality," Logsdon said. "There are a bunch of weird things about the genome, but are they directly related to putative asexuality or a consequence of other things?" Even though it's difficult to imagine how bdelloids' garbled chromosomes might pair for meiosis, they could have "some very unusual or infrequent process by which chromosomes pair and segregate," Logsdon said.

A NEW KIND OF SEX

So far, data from the bdelloid genome suggests that these creatures have survived by generating lots of genetic diversity through asexual means. But researchers haven't been able to prove it. Nor have they shown that this variation is enough to mimic sex. "It gets back to the question we keep asking theoretical biologists—how much sex is enough?" Mark Welch said. In other words, how much genetic scrambling does an organism need to do in order to mimic the benefits of sexual reproduction? To answer that question, scientists will need to measure genetic variability among a number of bdelloids and compare it to sexually reproducing populations.

Scientists don't yet have enough data to distinguish among various theories for why sex is so important—and it's possible that a number of these potential mechanisms contribute to bdelloids' lengthy survival. "One thing that gets theoretical biologists upset is to suggest many [theories] might be true," Mark Welch said. "But there's no particular biological reason that many theories can't be right."

Perhaps the more interesting question is how bdelloids succeeded when so many other asexual species have failed. Van Doninck is now exploring whether their striking ability to survive drought is the key to their long-term asexual existence. When a bdelloid dries out, its genome shatters into fragments, which the animal can stitch back together again once it's rehydrated.[6] It's possible that this remarkable DNA repair function allows bdelloids to scramble their chromosomes and to take up foreign DNA floating around in the environment, fixing these fragments into the genome as it recomposes. The result: a kind of supercharged genetic recombination without sex. Researchers are testing this idea by exposing bdelloids to rounds of radiation and desiccation and analyzing how the genome rearranges itself.

Early evidence hints that bdelloids could also assimilate DNA from other members of their own species. That's particularly significant because it would resemble traditional sex. "If they do genetic exchange among each other, then they have some kind of sex," Van Doninck said. That process wouldn't require meiosis, an essential component of sexual reproduction as currently defined. But in Van Doninck's view, it might be time to broaden the definition. Perhaps sex can be defined simply as a genetic exchange among members of the same species. Bdelloids might be the exception that changes the rule.

DID NEURONS EVOLVE TWICE?

Emily Singer

W hen Leonid Moroz, a neuroscientist at the Whitney Laboratory for Marine Bioscience in St. Augustine, Florida, first began studying comb jellies, he was puzzled. He knew the primitive sea creatures had nerve cells—responsible, among other things, for orchestrating the darting of their tentacles and the beat of their iridescent cilia. But those neurons appeared to be invisible. The dyes that scientists typically use to stain and study those cells simply didn't work. The comb jellies' neural anatomy was like nothing else he had ever encountered.

After years of study, he thinks he knows why. According to traditional evolutionary biology, neurons evolved just once, hundreds of millions of years ago, likely after sea sponges branched off the evolutionary tree. But Moroz thinks it happened twice—once in ancestors of comb jellies, which split off at around the same time as sea sponges, and once in the animals that gave rise to jellyfish and all subsequent animals, including us. He cites as evidence the fact that comb jellies have a relatively alien neural system, employing different chemicals and architecture from our own. "When we look at the genome and other information, we see not only different grammar but a different alphabet," Moroz said.

When Moroz proposed his theory, evolutionary biologists were skeptical. Neurons are the most complex cell type in existence, critics argued, capable of capturing information, making computations and executing decisions. Because they are so complicated, they are unlikely to have evolved twice.

But new support for Moroz's idea comes from recent genetic work suggesting that comb jellies are ancient—the first group to branch off the animal family tree. If true, that would bolster the chance that they evolved neurons on their own.

The debate has generated intense interest among evolutionary biologists. Moroz's work does not only call into question the origins of the brain and the evolutionary history of animals. It also challenges the deeply entrenched idea that evolution progresses steadily forward, building up complexity over time.

THE FIRST SPLIT

Somewhere in the neighborhood of 540 million years ago, the ocean was poised for an explosion of animal life. The common ancestor of all animals roamed the seas, ready to diversify into the rich panoply of fauna we see today.

Scientists have long assumed that sponges were the first to branch off the main trunk of the animal family tree. They're one of the simplest classes of animals, lacking specialized structures, such as nerves or a digestive system. Most rely on the ambient flow of water to collect food and remove waste.

Later, as is generally believed, the rest of the animal lineage split into comb jellies, also known as ctenophores (pronounced TEN-oh-fours); cnidarians (jellyfish, corals and anemones); very simple multicellular animals called placozoa; and eventually bilaterians, the branch that led to insects, humans and everything in between.

But sorting out the exact order in which the early animal branches split has been a notoriously thorny problem. We have little sense of what animals looked like so many millions of years ago because their soft bodies left little tangible evidence in rocks. "The fossil record is spotty," said Linda Holland, an evolutionary biologist at the Scripps Institution of Oceanography at the University of California, San Diego.

To make up for our inability to see into the past, scientists use the morphology (structure) and genetics of living animals to try to reconstruct the relationships of ancient ones. But in the case of comb jellies, the study of living animals presents serious challenges.

Little is known about comb jellies' basic biology. The animals are incredibly fragile, often falling to pieces once they're caught in a net. And it's difficult to raise them in captivity, making it nearly impossible to do the routine experiments that scientists might perform on other animals.

For a long time comb jellies were thought to be closely related to jellyfish. With their symmetrical body plans and gelatinous makeup, the two species outwardly resemble one another. Yet the animals swim and hunt differently—jellyfish have stinging tentacles, while comb jellies have sticky ones. And at the genome level, comb jellies are closer to sponges, which have no nervous system at all.

In comb jellies or in any other animal, an evolutionary analysis that relies on morphology might lead to one evolutionary tree, while one that uses genomic data, or even different kinds of genomic data, might lead to another. The discrepancies often spark heated debate in the field.

One such debate emerged in 2008, when Mark Martindale, now director of the Whitney Laboratory, Gonzalo Giribet, an evolutionary biologist at Harvard University, and collaborators published a study that analyzed gene sequences from 29 different animals.[1] After considering the genetic data, the researchers proposed a number of changes to the animal tree.

By far the most controversial of these changes was the suggestion that ctenophores should replace sponges as the earliest branch of animals. If evolution increases complexity over time, as biologists have traditionally believed, then an apparently simple organism like the sponge should predate a seemingly more complex organism like the comb jelly. Martindale and Giribet's genetic data suggested otherwise, but critics were dubious. "We were pretty much ridiculed by the entire scientific community," Martindale said.

Martindale and his collaborators needed to gather more evidence for their proposal. They convinced the National Institutes of Health to sequence the genome of a comb jelly, the sea walnut, which was published in *Science* in 2013.[2] Moroz and his collaborators published a second ctenophore genome, the sea gooseberry, in *Nature* in 2014. Both papers, which employed more extensive data and more sophisticated analysis methods than the 2008 effort, support the ctenophore-first tree. A third paper analyzing publicly available genome data and posted to the preprint server biorxiv.org in 2015 also supports the idea that comb jellies branched off first.[3]

In light of the new evidence, scientists are beginning to take the idea seriously, although many in the field say there isn't enough data to make any strong claims. This viewpoint was reflected in a flurry of review articles, many of which contended that comb jellies aren't really the oldest branch; they just appear to be.

Comb jellies have evolved more rapidly than the other ancient animal groups, meaning that their gene sequences changed quickly over time. This in turn means that the genetic analysis of their place in the evolutionary tree could be subject to a computational artifact called "long-branch attraction," a sort of glitch that can pull rapidly evolving organisms to the base of the tree. "Long-branched animal groups are often difficult to place," said Detlev Arendt, an evolutionary biologist at the European Molecular Biology Laboratory in Germany. "So far, the phylogenetic data is not really conclusive on where [comb jellies] belong."

Scientists hope that more data—including genomes of additional ctenophore species—will help resolve the deepest branches of the animal tree. And that, in turn, could have profound implications for our understanding of neurons and where they came from. "The branching order has a major influence on how we interpret the evolution of the nervous system," said

The Earliest Animals and the Origin of Neurons

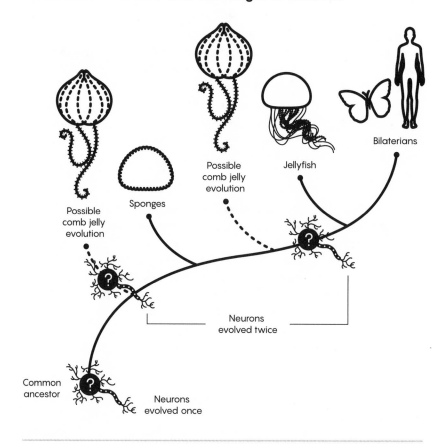

FIGURE 4.3

Scientists argue about the proper position of comb jellies on the animal evolutionary tree. Recent genetic evidence suggests that comb jelly ancestors were the first to split off from other animals. Other researchers maintain that sponges were likely first. Scientists also disagree over the origin of neurons. Were they present in the ancestor of all animals? Or did they evolve once in comb jellies and again in the ancestor to jellyfish and bilaterians?

Gáspár Jékely, a biologist at the Max Planck Institute for Developmental Biology in Germany.

Indeed, even those who agree that comb jellies came first disagree on the question of how neurons arose.

THE SPARK OF THOUGHT

The creation of neurons was a remarkable event in animal evolution. These cells can communicate—receiving, transmitting and processing information using a precise chemical and electrical language. Their power derives from the complex network they create. "A single neuron is like the sound of one hand clapping," Martindale said. "The whole idea is that you put a bunch of them together and they can do things that a few single cells cannot."

This level of complexity requires an unlikely confluence of evolutionary events. Mechanisms must arise that not only physically connect cells, but allow them to transmit and interpret signals. "The reason most people do not think that they could have evolved multiple times is the idea that neurons talk—specifically to other neurons," Martindale said.

That's what makes Moroz's proposal—that neurons evolved twice, once in comb jellies and once in other animals—so controversial.

According to Moroz's version of the evolutionary tree, animals started off with a common ancestor that had no neurons. Comb jellies then split off and went on to develop their strange brand of neurons. After that, ancestors of sponges and placozoans branched off. Like their ancestors, they lacked neurons. Rudimentary neurons, or protoneurons, then evolved for a second time in the ancestors of jellyfish and bilatarians, forming the basis of the nervous system found in all subsequent descendants, including humans. "In my opinion, it's simpler and more realistic that the common ancestor had no nervous system," Moroz said. (He thinks that even if comb jellies split off after sponges, they still evolved neurons independently.)

But some scientists who believe that ctenophores branched off first paint a different picture. They suggest that the common ancestor to all animals had a simple nervous system, which sponges subsequently lost. Comb jellies and the remaining branch, which includes our ancestors, the bilaterians, built on those protoneurons in different ways, developing increasingly sophisticated nervous systems.

"The ctenophores-first idea, if correct, suggests something really interesting going on," said Christopher Lowe, a biologist at the Hopkins Marine Station at Stanford University. "Both interpretations are profound." On the one hand, two independent origins of neurons would be surprising because it seems unlikely that the precise sequence of genetic accidents that created neurons could happen more than once. But it also seems unlikely that sponges would lose something as valuable as a neuron. "The only example we know from bilaterians where the nervous system was lost completely is in parasites," Lowe said.

The two possibilities reflect a classic conundrum for evolutionary biologists. "Did this animal lose something or not have it to begin with?" Holland said. In this particular case, "I find it's hard to take a stand," she said.

Evolution is rife with examples of both loss and parallel evolution. Some worms and other animals have shed regulatory molecules or developmental genes employed by the rest of the animal kingdom. "It's not unprecedented for important complements of genes to be lost in major animal lineages," Lowe said. Convergent evolution, in which natural selection produces two similar structures independently, is fairly common in nature. The retina, for example, evolved independently several times. "Different animals sometimes use extremely different toolkits to make morphologically similar neurons, circuits and brains," Moroz said. "Everyone accepts the eye case, but they think the brain or neuron only happened once."

Moroz's primary evidence for an independent origin of neurons in comb jellies comes from their unusual nervous systems. "The ctenophore's nervous system is dramatically different from any other nervous system," said Andrea Kohn, a molecular biologist who works with Moroz. Comb jellies appear to lack the commonly used chemical messengers that other animals have, such as serotonin, dopamine and acetylcholine. (They do use glutamate, a simple molecule that plays a major role in neuronal signaling in animals.) Instead, they have genes that are predicted to produce a slew of neural peptides, small proteins that can also act as chemical messengers. "No other animal except in this phylum has anything like that," Kohn said.

But critics question this assertion as well. Perhaps comb jellies really do have the genes for serotonin and other neural signaling molecules, but those genes have evolved beyond recognition, Arendt said. "It could just mean [that comb jellies] are highly specialized," he said.

Scientists on all sides of the debate say that it can only be answered with more data, and, more importantly, a better understanding of comb jelly biology. Even though they share some genes with model organisms, such as mice and fruit flies, it's unclear what those genes do in comb jellies. Nor do scientists understand their basic cell biology, like how comb jelly neurons communicate.

But the ongoing debate has sparked interest in ctenophores, and more researchers are studying their nervous systems, development and genes. "Moroz and collaborators have shined the light on this part of the tree, which is a good thing," Holland said. "We shouldn't ignore those guys down there."

V WHAT MAKES US HUMAN?

HOW HUMANS EVOLVED SUPERSIZE BRAINS

Ferris Jabr

T here it was, sitting on the mantelpiece, staring at her with hollow eyes and a naked grin. She could not stop staring back. It looked distinctly like the fossilized skull of an extinct baboon. That was the sort of thing Josephine Salmons was likely to know. At the time—1924—she was one of the only female students of anatomy attending the University of the Witwatersrand in South Africa. On this particular day she was visiting her friend Pat Izod, whose father managed a quarry company that had been excavating limestone near the town of Taung. Workers had unearthed numerous fossils during the excavation, and the Izods had kept this one as a memento. Salmons brought news of the skull to her professor, Raymond Dart, an anthropologist with a particular interest in the brain. He was incredulous. Very few primate fossils had been uncovered this far south in Africa. If the Taung site really housed such fossils, it would be an invaluable treasure trove. The next morning Salmons brought Dart the skull, and he could see that she was right: The skull was undeniably simian.

Dart promptly arranged to have other primate fossils from the Taung quarry sent to him. Later that year, as he was preparing to attend a close friend's wedding, he received a large crate. One of the specimens it contained was so mesmerizing that he nearly missed the ceremony. It came in two pieces: a natural endocast—the fossilized mold of the inner cranium, preserving the brain's topography—and its matching skeletal face, with eye sockets, nose, jaw and teeth all intact. Dart noticed right away that this was the fossil of an extinct ape, not a monkey. The teeth suggested that the individual had died at age 6 or so. The point where the spinal cord had joined the skull was too far forward for a knuckle walker, indicating bipedalism. And the endocast, which was a little too large for a nonhuman ape of that age, had surface features characteristic of a human brain. After further study, Dart reached a bold conclusion: this was the fossil of a previously unknown ancestor of modern humans—*Australopithecus africanus*, the "Man-Ape of South Africa."

At first, the greater scientific community lambasted Dart's proposal. If the Taung child, as the fossil was nicknamed, truly belonged to a hominin, surely it would have a far larger brain. Its cranium was a bit bigger than that of a chimpanzee, but not by much. Besides, it was generally believed that humans had evolved in Asia, not Africa. The "absurdly tiny" illustration accompanying Dart's 1925 *Nature* paper, and his initial possessiveness of the specimen, did not help matters.[1] Eventually, though, as prominent experts got to see the Taung child for themselves, and similar fossil discoveries came to light, attitudes began to change. By the 1950s, anthropologists had accepted that Taung was indeed a hominin and that an exceptionally large brain had not always been a distinguishing characteristic of humans. Dean Falk, a professor of anthropology at Florida State University and an expert on brain evolution, has called the Taung child "one of the most (if not *the* most) important hominin discoveries of the 20th century."

In subsequent decades, by uncovering and comparing other fossil skulls and endocasts, paleontologists documented one of the most dramatic transitions in human evolution. We might call it the Brain Boom. Humans, chimps and bonobos split from their last common ancestor between 6 and 8 million years ago. For the next few million years, the brains of early hominins did not grow much larger than those of our ape ancestors and cousins. Starting around 3 million years ago, however, the hominin brain began a massive expansion. By the time our species, *Homo sapiens*, emerged about 200,000 years ago, the human brain had swelled from about 350 grams to more than 1,300 grams. In that 3-million-year sprint, the human brain almost quadrupled the size its predecessors had attained over the previous 60 million years of primate evolution.

Fossils established the Brain Boom as fact. But they tell us next to nothing about how and why the human brain grew so large so quickly. There are plenty of theories, of course, especially regarding why: increasingly complex social networks, a culture built around tool use and collaboration, the challenge of adapting to a mercurial and often harsh climate—any or all of these evolutionary pressures could have selected for bigger brains.

Although these possibilities are fascinating, they are extremely difficult to test. In the last eight years, however, scientists have started to answer the "how" of human brain expansion—that is, the question of how the supersizing happened on a cellular level and how human physiology reconfigured itself to accommodate a dramatically enlarged and energy-guzzling brain. "It was all speculation up until now, but we finally have the tools to really get some traction," said Gregory Wray, an evolutionary biologist at Duke University. "What kinds of mutations occurred, and what did they do? We're starting to get answers and a deeper appreciation for just how complicated this process was."

WHAT MAKES THE HUMAN BRAIN SPECIAL

One scientist, in particular, has transformed the way researchers size up brains. Rather than fixating on mass or volume as a proxy for brainpower, she has focused on counting a brain's constituent parts.

In her laboratory at Vanderbilt University, Suzana Herculano-Houzel routinely dissolves brains into a soup of nuclei—cells' genetic control rooms. Each neuron has one nucleus. By tagging the nuclei with fluorescent molecules and measuring the glow, she can get a precise tally of individual brain cells. Using this method on a wide variety of mammalian brains, she has shown that, contrary to long-standing assumptions, larger mammalian brains do not always have more neurons, and the ones they do have are not always distributed in the same way.

Brain Size and Neuron Count

Cerebral cortex mass and neuron count for various mammals.

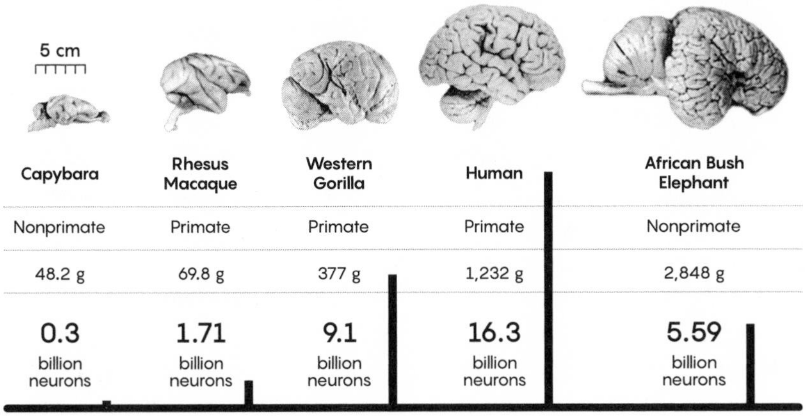

	Capybara	Rhesus Macaque	Western Gorilla	Human	African Bush Elephant
	Nonprimate	Primate	Primate	Primate	Nonprimate
	48.2 g	69.8 g	377 g	1,232 g	2,848 g
	0.3 billion neurons	1.71 billion neurons	9.1 billion neurons	16.3 billion neurons	5.59 billion neurons

FIGURE 5.1

When it comes to brains, size isn't everything. The human brain is much smaller than that of an elephant or whale. But there are far more neurons in a human's cerebral cortex than in the cortex of any other animal.
Source: BrainMuseum.org and Suzana Herculano-Houzel et al., "Brain Scaling in Mammalian Evolution as a Consequence of Concerted and Mosaic Changes in Numbers of Neurons and Average Neuronal Cell Size," *Frontiers in Neuroanatomy* 8 (2014): 77. https://doi.org/10.3389/fnana.2014.00077

The human brain has 86 billion neurons in all: 69 billion in the cerebellum, a dense lump at the back of the brain that helps orchestrate basic bodily functions and movement; 16 billion in the cerebral cortex, the brain's thick corona and the seat of our most sophisticated mental talents, such as self-awareness, language, problem solving and abstract thought; and 1 billion in the brain stem and its extensions into the core of the brain. In contrast, the elephant brain, which is three times the size of our own, has 251 billion neurons in its cerebellum, which helps manage a giant, versatile trunk, and only 5.6 billion in its cortex. Considering brain mass or volume alone masks these important distinctions.

Based on her studies, Herculano-Houzel has concluded that primates evolved a way to pack far more neurons into the cerebral cortex than other mammals did. The great apes are tiny compared to elephants and whales, yet their cortices are far denser: Orangutans and gorillas have 9 billion cortical neurons, and chimps have 6 billion. Of all the great apes, we have the largest brains, so we come out on top with our 16 billion neurons in the cortex. In fact, humans appear to have the most cortical neurons of any species on Earth. "That's the clearest difference between human and nonhuman brains," Herculano-Houzel said. It's all about the architecture, not just size.

The human brain is also unique in its unsurpassed gluttony. Although it makes up only 2 percent of body weight, the human brain consumes a whopping 20 percent of the body's total energy at rest. In contrast, the chimpanzee brain needs only half that. Researchers have long wondered how the human body adapted to sustain such a uniquely ravenous organ. In 1995, the anthropologist Leslie Aiello and the evolutionary biologist Peter Wheeler proposed the "expensive tissue hypothesis" as a possible answer. The underlying logic is straightforward: Human brain evolution likely required a metabolic trade-off. In order for the brain to grow, other organs, namely the gut, had to shrink, and energy that would typically have gone to the latter was redirected to the former. For evidence, they pointed to data showing that primates with larger brains have smaller intestines.

A few years later, the anthropologist Richard Wrangham built on this idea, arguing that the invention of cooking was crucial to human brain evolution. Soft, cooked foods are much easier to digest than tough raw ones, yielding more calories for less gastrointestinal work. Perhaps, then, learning to cook permitted a bloating of the human brain at the expense of the gut. Other researchers have proposed that similar trade-offs might have occurred between brain and muscle, given how much stronger chimps are than humans.

Collectively, these hypotheses and observations of modern anatomy are compelling. But they are based on the echoes of biological changes that

Brain Density

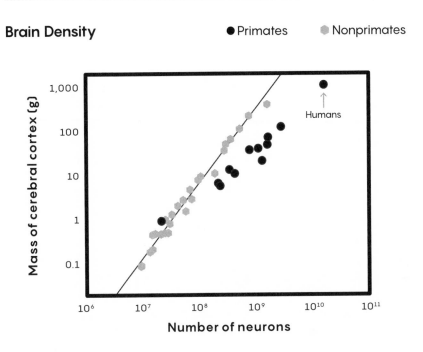

FIGURE 5.2
How does the number of neurons in the cerebral cortex vary with the size of that part of the brain? Different scaling rules apply. In rodents, a 10-fold increase in the number of cortical neurons leads to a 50-fold increase in the size of the cortex. In primates, by contrast, the same neural increase leads to only a 10-fold increase in cortex size—a far more economical relationship. Source: Herculano-Houzel et al. (2014)

are thought to have occurred millions of years ago. To be certain of what happened, to pinpoint the physiological adaptations that made the brain's evolutionary growth spurt possible, we will have to dive deeper than flesh, into our very genome.

HOW GENES BUILD THE BRAIN

About a decade ago, Wray and his colleagues began to investigate a family of genes that influence the movement of glucose into cells to be used as energy. One member of the gene family is especially active in brain tissue, whereas another is most active in muscle. If the size of the human brain required a metabolic trade-off between brain tissue and muscle, then these genes should behave differently in humans and chimpanzees.

Wray and his team collected brain, muscle and liver samples from deceased humans and chimpanzees and attempted to measure gene activity in each sample. When a cell "expresses" a gene, it translates the DNA first into a signature messenger RNA (mRNA) sequence and subsequently into a chain of amino acids that forms a protein. Varying levels of distinct mRNAs can therefore provide a snapshot of gene activity in a particular type of tissue.

Wray's team extracted mRNA from the tissues and amplified it many times over in the lab in order to measure the relative abundance of different mRNAs. They found that the brain-centric glucose-transporting gene was 3.2 times more active in human brain tissue than in the chimp brain, whereas the muscle-centric gene was 1.6 times more active in chimp muscle than in human muscle. Yet the two genes behaved similarly in the liver of both species.

Given that the human and chimp gene sequences are nearly identical, something else must explain their variable behavior. Wray and his colleagues found some intriguing differences between the genes' corresponding regulatory sequences—stretches of DNA that stimulate or stifle gene activity. In humans, but not in chimps, the regulatory sequences for the muscle and brain-focused glucose-transporting genes had accumulated more mutations than would be expected by chance alone, indicating that these regions had undergone accelerated evolution. In other words, there was a strong evolutionary pressure to modify the human regulatory regions in a way that sapped energy from muscle and channeled it to the brain. Genes had corroborated the expensive tissue hypothesis in a way fossils never could.

In 2014, the computational biologist Kasia Bozek, who now works at the Okinawa Institute for Science and Technology in Japan, published a similar study that examined metabolism from a different angle. In addition to looking at gene expression, Bozek and her colleagues analyzed levels of metabolites, a diverse group of small molecules that includes sugars, nucleic acids and neurotransmitters. Many metabolites are either necessary for metabolism or produced by it. Different organs have distinct metabolite profiles, depending on what they do and how much energy they require. In general, metabolite levels in the organs of closely related species are more in sync than levels between distantly related species. Bozek found that the metabolite profiles of human and chimp kidneys, for example, were pretty similar. But the variation between chimp and human brain metabolite levels was four times higher than would be expected based on a typical rate of evolution; muscle metabolites differed from the expected levels by a factor of seven. "A single gene can probably regulate a lot of metabolites," Bozek said. "So even if the difference is not huge at the gene level, you could get a big difference in the metabolite levels."

Bozek and her colleagues then pitted 42 humans, including college basketball players and professional rock climbers, against chimpanzees and macaques in a test of strength. All of the primates had to pull a sliding shelf saddled with weights toward themselves. Accounting for body size and weight, the chimps and macaques were twice as strong as the humans. It's not entirely clear why, but it is possible that our primate cousins get more power out of their muscles than we get out of ours because they feed their muscles more energy. "Compared to other primates, we lost muscle power in favor of sparing energy for our brains," Bozek said. "It doesn't mean that our muscles are inherently weaker. We might just have a different metabolism."

Meanwhile, Wray had turned to his Duke colleague Debra Silver, an expert in embryonic brain development, to embark on a pioneering experiment. Not only were they going to identify relevant genetic mutations from our brain's evolutionary past, they were also going to weave those mutations into the genomes of lab mice and observe the consequences. "This is something no one had attempted before," Silver said.

The researchers began by scanning a database of human accelerated regions (HARs); these regulatory DNA sequences are common to all vertebrates but have rapidly mutated in humans. They decided to focus on HARE5, which seemed to control genes that orchestrate brain development. The human version of HARE5 differs from its chimp correlate by 16 DNA letters. Silver and Wray introduced the chimpanzee copy of HARE5 into one group of mice and the human edition into a separate group. They then observed how the embryonic mice brains grew.

After nine days of development, mice embryos begin to form a cortex, the outer wrinkly layer of the brain associated with the most sophisticated mental talents. On day 10, the human version of HARE5 was much more active in the budding mice brains than the chimp copy, ultimately producing a brain that was 12 percent larger. Further tests revealed that HARE5 shortened the time required for certain embryonic brain cells to divide and multiply from 12 hours to nine. Mice with the human HARE5 were creating new neurons more rapidly.

"This sort of study would have been impossible to do 10 years ago when we didn't have the full genome sequences," Silver said. "It's really exciting." But she also stressed that it will take a great deal more research to fully answer how the human brain blew up. "It's a mistake to think we can explain brain size with just one or two mutations. I think that is dead wrong. We have probably acquired many little changes that are in some ways coopting the developmental rules."

Wray concurs: "It wasn't just a couple mutations and—bam!—you get a bigger brain. As we learn more about the changes between human and chimp brains, we realize there will be lots and lots of genes involved, each contributing a piece to that. The door is now open to get in there and really start understanding. The brain is modified in so many subtle and nonobvious ways."

BRAIN AND BODY

Although the mechanics of the human brain's expansion have long been mysterious, its importance has rarely been questioned. Again and again, researchers have cited the evolutionary surge in human brain size as the key reason for our exceptionally high degree of intelligence compared to other animals. As recent research on whale and elephant brains makes clear, size is not everything, but it certainly counts for something. The reason we have so many more cortical neurons than our great-ape cousins is not that we have denser brains, but rather that we evolved ways to support brains that are large enough to accommodate all those extra cells.

There's a danger, though, in becoming too enamored with our own big heads. Yes, a large brain packed with neurons is essential to what we consider high intelligence. But it's not sufficient. Consider, for a moment, what the world would be like if dolphins had hands. Dolphins are impressively brainy. They have demonstrated self-awareness, cooperation, planning and the rudiments of language and grammar. Compared to apes, though, they are severely limited in their ability to manipulate the world's raw materials. Dolphins will never enter the Stone Age; flippers cannot finesse.

Similarly, we know that chimps and bonobos can understand human language and even form simple sentences with touch-screen keyboards, but their vocal tracts are inadequate for producing the distinct series of sounds required for speech. Conversely, some birds have the right vocal anatomy to flawlessly mimic human speech, but their brains are not large enough or wired in the right way to master complex language.

No matter how large the human brain grew, or how much energy we lavished upon it, it would have been useless without the right body. Three particularly crucial adaptations worked in tandem with our burgeoning brain to dramatically increase our overall intelligence: bipedalism, which freed up our hands for tool making, fire building and hunting; manual dexterity surpassing that of any other animal; and a vocal tract that allowed us to speak and sing. Human intelligence, then, cannot be traced to a single organ, no matter how large; it emerged from a serendipitous confluence of adaptations throughout the body. Despite our ongoing obsession with the size of our noggins, the fact is that our intelligence has always been so much bigger than our brain.

NEW EVIDENCE FOR THE NECESSITY OF LONELINESS

Emily Singer

As social animals, we depend on others for survival. Our communities provide mutual aid and protection, helping humanity to endure and thrive. "We have survived as a species not because we're fast or strong or have natural weapons in our fingertips, but because of social protection," said John Cacioppo, then director of the Center for Cognitive and Social Neuroscience at the University of Chicago, in 2016. Early humans, for example, could take down large mammals only by hunting in groups. "Our strength is our ability to communicate and work together," he said.

But how did these powerful communities come to exist in the first place? Cacioppo proposed that the root of social ties lies in their opposite—loneliness. According to his theory, the pain of being alone motivates us to seek the safety of companionship, which in turn benefits the species by encouraging group cooperation and protection. Loneliness persists because it provides an essential evolutionary benefit for social animals. Like thirst, hunger or pain, loneliness is an aversive state that animals seek to resolve, improving their long-term survival.

If Cacioppo's theory is correct, then there must be an intrinsic biological mechanism that compels isolated animals to seek out companionship. Something in our brains must make it feel bad to be alone and bring relief when we're with others. Researchers at the Massachusetts Institute of Technology think they've found the source of that motivation in a group of little-studied neurons in a part of the brain called the dorsal raphe nucleus. Stimulating these neurons drives isolated mice to find friends, according to research published in 2016 in the journal *Cell*.[1] The finding provides critical support to Cacioppo's theory and illuminates a deep connection that links specific structures in the brain to social behavior.

The study—the first to link specific neurons to loneliness—is part of a growing effort to map out the genetics of social behavior and its underpinnings in the brain. "Over the last roughly 15 years, there has been a tremendous increase in the desire to understand the basis of social behavior, including caring for others, social rejection, bullying, deceit and so forth," said Patricia Churchland, a philosopher at the University of California, San Diego, who studies the brain and social behavior. "I think we have a good idea for the evolutionary basis for caring and sharing and mutual defense, but the brain mechanisms are bound to be very complex."

Together, Cacioppo's work and the new findings from MIT are helping to move loneliness from the realm of psychology and literature to biology. "I think the bigger picture is not to understand why loneliness is painful but rather how our brain is set up to move us out of that lonely state," said Steve Cole, a genomics researcher at the University of California, Los Angeles. "Instead of thinking about loneliness, we could think about social affinity."

SOCIAL CREATURES

Gillian Matthews stumbled across the loneliness neurons by accident. In 2012 she was a graduate student at Imperial College London who had been studying how cocaine changes the brain in mice. She would give the animals a dose of the drug, place each one alone in a cage, and then examine a specific set of its neurons the next day. She did the same for a control group of mice, injecting them with saline instead of cocaine.

When Matthews returned to her mice 24 hours after dosing them, she expected to see changes in their brain cells, a strengthening of neuronal connections that might help explain why cocaine is so addictive. To her surprise, both the drug-treated mice and the control mice showed the same changes in neuronal wiring. Overnight, the neural connections onto a certain set of cells had grown stronger, regardless of whether the animals were given drugs or not. "We first thought there was something wrong, that we had mixed up our procedure," said Matthews, who is now a postdoctoral researcher at MIT.

The brain cells she was interested in produce dopamine, a brain chemical typically associated with pleasurable things. Dopamine surges when we eat, have sex or use drugs. But it does more than simply signal pleasure. The brain's dopamine systems may be set up to drive the search for what we desire. "It's not what happens after you get what you want, it's what keeps you searching for something," Cole said.

The researchers focused on dopamine neurons in a brain region called the dorsal raphe nucleus, best known for its link to depression. (This may not be a coincidence—loneliness is a strong risk factor for depression.) Most of the neurons that reside there produce serotonin, the chemical messenger that drugs such as Prozac act on. Dopamine-producing cells make up roughly 25 percent of the region and have historically been difficult to study on their own, so scientists know little about what they do.

Matthews speculated that other environmental factors during the experiment might have triggered the changes. She tested to see if simply moving mice to new cages altered the dopamine neurons, but that couldn't explain the effect. Ultimately, Matthews and her colleague Kay Tye realized that these brain cells were responding not to the drug but to the 24 hours of isolation. "Maybe these neurons are relaying the experience of loneliness," Matthews said.

Mice, like humans, are social creatures that generally prefer to live in groups.[2] Isolate a mouse from its cage mates, and once confinement ends it will spend more time interacting with other mice, to a much greater extent than if it had been with its mates all along.[3]

To better understand the role the dorsal raphe neurons play in loneliness, the researchers genetically engineered the dopamine cells to respond to certain wavelengths of light, a technique known as optogenetics. They could then artificially stimulate or silence the cells by exposing them to light.

Stimulating the dopamine neurons seemed to make the mice feel bad. Mice actively avoided stimulation if given the choice, just as they might avoid physical pain. Moreover, the animals appeared to enter a state of loneliness—they acted like they had been alone, spending more time with other mice.

"I think this reveals something about how our brains may be wired to make us innately social creatures and protect us from the detrimental effects of loneliness," Matthews said.

SPECTRUM OF LONELINESS

Cacioppo first formally proposed his evolutionary theory of loneliness more than a decade ago.[4] Strong support comes from the fact that our sensitivity to loneliness is heritable, like height or risk of diabetes—about 50 percent of an individual's level of loneliness can be tied to their genes. "If it was really bad, it would have been bred out, so it must be adaptive," said Louise Hawkley, a psychologist at NORC at the University of Chicago, who has collaborated with Cacioppo in the past. The evolutionary theory for loneliness "forms a very coherent story about how loneliness might have come to exist," she said.

Indeed, like diabetes, people have varying degrees of susceptibility to loneliness. "What's being inherited is not loneliness, it's the painfulness of the disconnection," said Cacioppo, who in 2016 was trying to nail down the specific genes linked to loneliness with studies of tens of thousands of people.

In evolutionary terms, it's helpful for a population to have some variability in this trait. Some members of a community would be "so pained by disconnection that they are willing to defend their village," Cacioppo said. "Others are willing to go out and explore but hopefully still have enough of a connection to come back and share what they found."

Mice also show this variability. In Matthews' experiments, the most dominant mice—those that win in fights against their cage mates and have priority access to food and other resources—show the strongest reaction to having their loneliness neurons stimulated. At those times, the highest ranking animals search out companionship more fervently than animals on the lowest rungs of the social ladder. These mice also avoid stimulation of the loneliness neurons more avidly than the lower ranking members, suggesting that the dominant mice find it more unpleasant. The lowest ranked mice, in contrast, didn't seem to mind being alone. Perhaps they enjoyed isolation, being free of their harassers.

"It's extremely complex—they see a lot of variability just in rodents," Churchland said. "I think that's really very striking."

Tye and Matthews' findings suggest that these dorsal raphe nucleus neurons help to resolve the disconnect between the level of social connection the animal has and the level it wants. Imagine loneliness as a desire for ice cream—some animals love ice cream and some don't. The dopamine neurons drive the ice cream lovers to seek out the dessert but have little effect on everyone else. "We think that the [dorsal raphe nucleus] neurons are somehow tapping into that subjective social experience of the mouse, and only producing a significant effect on the behavior of mice who perhaps previously valued their social connections, rather than those who did not," Matthews said.

The varying reactions suggest two intriguing possibilities: either neural wiring determines social rank, or social rank influences how these neurons get wired. Perhaps some animals are wired from birth to crave social contact. These animals then seek out others and become aggressive as they try to maintain their position in the group, eventually attaining top status. Alternatively, certain mice may start out with an aggressive personality, picking on other animals in their group. The brain wiring in these animals might change as a result, driving the mice to seek out others to bully. Tye and Matthews are planning additional experiments to distinguish those two possibilities.

Cacioppo said he almost "fell over" when he saw Tye and Matthews' results. He had done extensive research on loneliness in humans, using brain imaging to identify parts of the brain that are active when people feel lonely. But brain imaging has a coarse resolution and can't analyze specific cell types like Tye and Matthews did in mice.

Tye and Matthews' research helps to reframe loneliness from a state of profound despair to a motivational force encoded in our biology. "Instead of focusing on the aversive state of being alone, this study looks at how social contact gets rewarded in the nervous system," Cole said. "Then loneliness becomes understandable as a lack of reward."

HOW NEANDERTHAL DNA HELPS HUMANITY

Emily Singer

E arly human history was a promiscuous affair. As modern humans began
to spread out of Africa roughly 50,000 years ago, they encountered
other species that looked remarkably like them—the Neanderthals and
Denisovans, two groups of archaic humans that shared an ancestor with
us roughly 600,000 years earlier. This motley mix of humans coexisted in
Europe for at least 2,500 years, and we now know that they interbred, leaving a lasting legacy in our DNA.[1,2] The DNA of non-Africans is made up
of roughly 1 to 2 percent Neanderthal DNA, and some Asian and Oceanic
island populations have as much as 6 percent Denisovan DNA.[3]

Over the last few years, scientists have dug deeper into the Neanderthal
and Denisovan sections of our genomes and come to a surprising conclusion. Certain Neanderthal and Denisovan genes seem to have swept
through the modern human population—one variant, for example, is present in 70 percent of Europeans—suggesting that these genes brought great
advantage to their bearers and spread rapidly.

"In some spots of our genome, we are more Neanderthal than human,"
said Joshua Akey, a geneticist now at Princeton University. "It seems pretty
clear that at least some of the sequences we inherited from archaic hominins were adaptive, that they helped us survive and reproduce."

But what, exactly, do these fragments of Neanderthal and Denisovan
DNA do? What survival advantage did they confer on our ancestors? Scientists are starting to pick up hints. Some of these genes are tied to our
immune system, to our skin and hair, and perhaps to our metabolism and
tolerance for cold weather, all of which might have helped emigrating
humans survive in new lands.

"What allowed us to survive came from other species," said Rasmus
Nielsen, an evolutionary biologist at the University of California, Berkeley.
"It's not just noise, it's a very important substantial part of who we are."

THE NEANDERTHAL WITHIN

The Tibetan plateau is a vast stretch of high-altitude real estate isolated by massive mountain ranges. The scant oxygen at 14,000 feet—roughly 40 percent lower than the concentrations at sea level—makes it a harsh environment. People who move there suffer higher rates of miscarriage, blood clots and stroke on account of the extra red blood cells their bodies produce to feed oxygen-starved tissue. Native Tibetans, however, manage just fine. Despite the meager air, they don't make as many red blood cells as the rest of us would at those altitudes, which helps to protect their health.

In 2010, scientists discovered that Tibetans owe their tolerance of low oxygen levels in part to an unusual variant in a gene known as *EPAS1*.[4] About 90 percent of the Tibetan population and a smattering of Han Chinese (who share a recent ancestor with Tibetans) carry the high-altitude variant. But it's completely absent from a database of 1,000 human genomes from other populations.

In 2014, Nielsen and colleagues found that Tibetans or their ancestors likely acquired the unusual DNA sequence from Denisovans, a group of early humans first described in 2010 that are more closely related to Neanderthals than to us.[5,6] The unique gene then flourished in those who lived at high altitudes and faded away in descendants who colonized less harsh environments. "That's one of the most clear-cut examples of how [interbreeding] can lead to adaptation," said Sriram Sankararaman, a geneticist and computer scientist at the University of California, Los Angeles.

The idea that closely related species can benefit from interbreeding, known in evolutionary terms as adaptive introgression, is not a new one. As a species expands into a new territory, it grapples with a whole new set of challenges—different climate, food, predators and pathogens. Species can adapt through traditional natural selection, in which spontaneous mutations that happen to be helpful gradually spread through the population. But such mutations strike rarely, making it a very slow process. A more expedient option is to mate with species that have already adapted to the region and co-opt some of their helpful DNA. (Species are traditionally defined by their inability to mate with one another, but closely related species often interbreed.)

This phenomenon has been well documented in a number of species, including mice that adopted other species' tolerance to pesticides and butterflies that appropriated other species' wing patterning.[7] But it was difficult to study adaptive introgression in humans until the first Neanderthal genome was sequenced in 2010, providing scientists with hominin DNA to compare to our own.

Neanderthals and Denisovans would have been a good source of helpful DNA for our ancestors. They had lived in Europe and Asia for hundreds of thousands of years—enough time to adjust to the cold climate, weak sun and local microbes. "What better way to quickly adapt than to pick up a gene variant from a population that had probably already been there for 300,000 years?" Akey said. Indeed, the Neanderthal and Denisovan genes with the greatest signs of selection in the modern human genome "largely have to do with how humans interact with the environment," he said.

To find these adaptive segments, scientists search the genomes of contemporary humans for regions of archaic DNA that are either more common or longer than expected. Over time, useless pieces of Neanderthal DNA—those that don't help the carrier—are likely to be lost. And long sections of archaic DNA are likely to be split into smaller segments unless there is selective pressure to keep them intact.

In 2014, two groups, one led by Akey and the other by David Reich, a geneticist at Harvard Medical School, independently published genetic maps that charted where in our genomes Neanderthal DNA is most likely to be found.[8,9] To Akey's surprise, both maps found that the most common adaptive Neanderthal-derived genes are those linked to skin and hair growth. One of the most striking examples is a gene called *BNC2*, which is linked to skin pigmentation and freckling in Europeans. Nearly 70 percent of Europeans carry the Neanderthal version.

Scientists surmise that *BNC2* and other skin genes helped modern humans adapt to northern climates, but it's not clear exactly how. Skin can have many functions, any one of which might have been helpful. "Maybe skin pigmentation, or wound healing, or pathogen defense, or how much water loss you have in an environment, making you more or less susceptible to dehydration," Akey said. "So many potential things could be driving this—we don't know what differences were most important."

SURVEILLANCE SYSTEM

One of the deadliest foes that modern humans had to fight as they ventured into new territories was also the smallest—novel infectious diseases for which they had no immunity. "Pathogens are one of the strongest selective forces out there," said Janet Kelso, a bioinformatician at the Max Planck Institute for Evolutionary Anthropology in Leipzig, Germany.

In 2016, Kelso and collaborators identified a large stretch of Neanderthal DNA—143,000 DNA base-pairs long—that may have played a key role in helping modern humans fight off disease. The region spans three different genes that are part of the innate immune system, a molecular surveillance system that forms the first line of defense against pathogens. These genes produce proteins called toll-like receptors, which help immune cells detect foreign invaders and trigger the immune system to attack.[10]

Modern humans can have several different versions of this stretch of DNA. But at least three of the variants appear to have come from archaic humans—two from Neanderthals and one from Denisovans. To figure out what those variants do, Kelso's team scoured public databases housing reams of genomic and health data. They found that people carrying one of the Neanderthal variants are less likely to be infected with *H. pylori*, a microbe that causes ulcers, but more likely to suffer from common allergies such as hay fever.

Kelso speculates that this variant might have boosted early humans' resistance to different kinds of bacteria. That would have helped modern humans as they colonized new territories. Yet this added resistance came at a price. "The trade-off for that was a more sensitive immune system that was more sensitive to nonpathogenic allergens," said Kelso. But she was careful to point out that this is just a theory. "At this point, we can hypothesize a lot, but we don't know exactly how this is working."

Most of the Neanderthal and Denisovan genes found in the modern genome are more mysterious. Scientists have only a vague idea of what these genes do, let alone how the Neanderthal or Denisovan version might have helped our ancestors. "It's important to understand the biology of these genes better, to understand what selective pressures were driving the changes we see in present-day populations," Akey said.

A number of studies like Kelso's are now under way, trying to link Neanderthal and Denisovan variants frequently found in contemporary humans with specific traits, such as body-fat distribution, metabolism or other factors. One study of roughly 28,000 people of European descent, published in *Science* in February 2016, matched archaic gene variants with data from electronic health records.[11] Overall, Neanderthal variants are linked to higher risk of neurological and psychiatric disorders and lower risk of digestive problems. (That study didn't focus on adaptive DNA, so it's unclear how the segments of archaic DNA that show signs of selection affect us today.)

At present, much of the data available for such studies is weighted toward medical problems—most of these databases were designed to find genes linked to diseases such as diabetes or schizophrenia. But a few, such as the UK Biobank, are much broader, storing information on participants' vision, cognitive test scores, mental health assessments, lung capacity and fitness. Direct-to-consumer genetics companies also have large, diverse data sets. For example, 23andMe analyzes users' genetics for clues about ancestry, health risk and other sometimes bizarre traits, such as whether they have a sweet tooth or a unibrow.

Of course, not all the DNA we got from Neanderthals and Denisovans was good. The majority was probably detrimental. Indeed, we tend to have less Neanderthal DNA near genes, suggesting that it was weeded out by natural selection over time. Researchers are very interested in these parts of our genomes where archaic DNA is conspicuously absent. "There are some really big places in the genome with no Neanderthal or Denisovan ancestry as far as we can see—some process is purging the archaic material from these regions," Sankararaman said. "Perhaps they are functionally important for modern humans."

THE NEUROSCIENCE BEHIND BAD DECISIONS

Emily Singer

Humans often make bad decisions. If you like Snickers more than Milky Way, it seems obvious which candy bar you'd pick, given a choice of the two. Traditional economic models follow this logical intuition, suggesting that people assign a value to each choice—say, Snickers: 10, Milky Way: 5—and select the top scorer. But our decision-making system is subject to glitches.

In a 2016 experiment, Paul Glimcher, a neuroscientist at New York University, and collaborators asked people to choose among a variety of candy bars, including their favorite—say, a Snickers. If offered a Snickers, a Milky Way and an Almond Joy, participants would always choose the Snickers. But if they were offered 20 candy bars, including a Snickers, the choice became less clear. They would sometimes pick something other than the Snickers, even though it was still their favorite. When Glimcher would remove all the choices except the Snickers and the selected candy, participants would wonder why they hadn't chosen their favorite.

Economists have spent more than 50 years cataloging irrational choices like these. Nobel Prizes have been earned; millions of copies of *Freakonomics* have been sold. But economists still aren't sure why they happen. "There had been a real cottage industry in how to explain them and lots of attempts to make them go away," said Eric Johnson, a psychologist and co-director of the Center for Decision Sciences at Columbia University. But none of the half-dozen or so explanations are clear winners, he said.

In the last 15 to 20 years, neuroscientists have begun to peer directly into the brain in search of answers. "Knowing something about how information is represented in the brain and the computational principles of the brain helps you understand why people make decisions how they do," said Angela Yu, a theoretical neuroscientist at the University of California, San Diego.

Glimcher is using both the brain and behavior to try to explain our irrationality. He has combined results from studies like the candy bar experiment with neuroscience data—measurements of electrical activity in the brains of animals as they make decisions—to develop a theory of how we make decisions and why that can lead to mistakes.

Glimcher has been one of the driving forces in the still young field of neuroeconomics. His theory merges far-reaching research in brain activity, neuronal networks, fMRI and human behavior. "He's famous for arguing that neuroscience and economics should be brought together," said Nathaniel Daw, a neuroscientist at Princeton University. One of Glimcher's most important contributions, Daw said, has been figuring out how to quantify abstract notions such as value and study them in the lab.

Glimcher and his co-authors—Kenway Louie, also of NYU, and Ryan Webb of the University of Toronto—argue that their neuroscience-based model outperforms standard economic theory at explaining how people behave when faced with lots of choices.[1] "The neural model, described in biology and tested in neurons, works well to describe something economists couldn't explain," Glimcher said.

At the core of the model lies the brain's insatiable appetite. The brain is the most metabolically expensive tissue in the body. It consumes 20 percent of our energy despite taking up only 2 to 3 percent of our mass. Because neurons are so energy-hungry, the brain is a battleground where precision and efficiency are opponents. Glimcher argues that the costs of boosting our decision-making precision outweigh the benefits. Thus we're left to be confounded by the choices of the modern American cereal aisle.

Glimcher's proposal has attracted interest from both economists and neuroscientists, but not everyone is sold. "I think it's exciting but at this point remains a hypothesis," said Camillo Padoa-Schioppa, a neuroscientist at Washington University in St. Louis. Neuroeconomics is still a young field; scientists don't even agree on what part of the brain makes decisions, let alone how.

So far, Glimcher has shown that his theory works under specific conditions, like those of the candy bar experiment. He aims to expand that range, searching for other *Freakonomics*-esque mistakes and using them to test his model. "We are aiming for a grand unified theory of choice," he said.

DIVIDE AND CONQUER

The brain is a power-hungry organ; neurons are constantly sending each other information in the form of electrical pulses, known as spikes or action potentials. Just as with an electrical burst, prepping and firing these signals take a lot of energy.

In the 1960s, scientists proposed that the brain dealt with this challenge by encoding information as efficiently as possible, a model called the efficient coding hypothesis. It predicts that neurons will encode data using the fewest possible spikes, just as communication networks strive to transmit information in the fewest bits.

In the late 1990s and early 2000s, scientists showed that this principle is indeed at work in the visual system.[2,3] The brain efficiently encodes the visual world by ignoring predictable information and focusing on the surprising stuff. If one part of a wall is yellow, chances are the rest is also yellow, and neurons can gloss over the details of that section.[4] But a giant red splotch on the wall is unexpected, and neurons will pay special attention to it.

Glimcher proposes that the brain's decision-making machinery works the same way. Imagine a simple decision-making scenario: a monkey choosing between two cups of juice. For simplicity's sake, assume the monkey's brain represents each choice with a single neuron. The more attractive the choice is, the faster the neuron fires. The monkey then compares neuron-firing rates to make his selection.

The first thing the experimenter does is present the monkey with an easy choice: a teaspoon of yummy juice versus an entire jug. The teaspoon neuron might fire one spike per second while the jug neuron fires 100 spikes per second. In that case, it's easy to tell the difference between the two options; one neuron sounds like a ticking clock, the other the beating wings of a dragonfly.

The situation gets muddled when the monkey is then offered the choice between a full jug of juice and one that's nearly full. A neuron might represent that newest offer with 80 spikes per second. It's much more challenging for the monkey to distinguish between a neuron firing 80 spikes per second and 100 spikes per second. That's like telling the difference between the dragonfly's flutter and the hum of a locust.

Glimcher proposes that the brain avoids this problem by recalibrating the scale to best represent the new choice. The neuron representing the almost-full jug—now the worst of the two choices—scales down to a much lower firing rate. Once again it's easy for the monkey to differentiate between the two choices.

Glimcher's model, based on an earlier model known as divisive normalization, spells out the math behind this recalibration process.[5] It proposes that neurons can send more efficient messages if they encode in their sequence of spikes only the relative differences among the choices. "Choice sets have a lot of shared information; they are not random and independent," Glimcher said. "Normalization is sucking out redundant information so that the information coming out is as relevant as possible, wasting as little energy as possible." He notes that engineers, who are used to working with adaptive systems, aren't surprised by this idea. But people who study choice often are.

According to Daw, "What's great about divisive normalization is that it takes these principles we know from vision and applies them to value in ways that make sense but are out of the box."

The juice example above is theoretical, but Glimcher and collaborators have recorded electrical activity from monkeys' brains as they make different kinds of choices. These studies show that decision-making neurons behave as the model predicts.[6,7] If scientists increase the value of one choice, the equivalent of swapping out a so-so Milky Way with a delicious Snickers, the neurons representing that choice increase their firing rate. (Scientists had already known about this pattern.)

If you increase the value of the other choices—king-size the non-Snickers options, which decreases the relative value of the Snickers—the model predicts that its firing rate should go down. Glimcher and collaborators have shown that neurons in part of the brain called the parietal cortex do indeed behave this way, adding physiological support for the model. "The divisive normalization function did a superb job of describing the data in all conditions," Glimcher said. "It supports the idea that neurons are doing something identical to, or darn close to, divisive normalization."

The system works well most of the time. But just like the temporary blindness we experience when exiting a dark movie theater into bright sunlight, our decision-making machinery can sometimes be overwhelmed. That may be particularly true with the staggering variety of choices we're often faced with in the modern world. Glimcher and collaborators use these types of mistakes to test their model. The researchers are examining whether these same algorithms can predict human error in other scenarios in which people tend to make poor choices.

ECONOMIC INSURGENCY

Neuroeconomics is still a young field, filled with questions and controversy. Glimcher isn't the only neuroscientist to have found signs of economic value in the brain. Scientists have measured these neural signatures in different brain regions, using both noninvasive brain imaging in humans and direct brain recordings in animals. But researchers disagree over which part of the brain makes the actual decision. What part of the brain calculates that the Snickers bar rates higher than the Milky Way? "There is no single accepted concept of where and how decisions—the comparison of values— are made," Padoa-Schioppa said.

Glimcher's neural recording experiments took place in the parietal cortex, but Padoa-Schioppa is "skeptical that the parietal cortex has anything to do with economic decisions." Damaging the parietal cortex doesn't impair value-based choices, he said, while damaging the frontal lobe does. For that reason, Padoa-Schioppa is somewhat dubious of Glimcher's model. When it comes to a neuroscience-based model of choice, "at this point, no one has a compelling theory," Padoa-Schioppa said.

Other scientists like the general concept of divisive normalization but suggest it can be refined to account for more complex aspects of human decision-making. Yu, for example, says it works well for simple decisions but may falter under more sophisticated conditions. "The divisive normalization model does make sense, but the experimental setting in which they were probing decision-making is very simplistic," Yu said. "To account for the broader array of phenomena in human decision-making, we need to augment the model and look at more complex decision-making scenarios."

The divisive normalization framework emerged from work in the visual system. Yu suggests that applying it to decision-making is more complex. Scientists know a lot about the information that the visual system is trying to encode: a two-dimensional scene painted in color, light and shadow. Natural scenes conform to a set of general, easy-to-calculate properties that the brain can use to filter out redundant information. In simple terms, if one pixel is green, its neighboring pixels are more likely to be green than red.

But the decision-making system operates under more complex constraints and has to consider many different types of information. For example, a person might choose which house to buy depending on its location, size or style. But the relative importance of each of these factors, as well as their optimal value—city or suburbs, Victorian or modern—is fundamentally subjective. It varies from person to person and may even change for an individual depending on their stage of life. "There is not one simple, easy-to-measure mathematical quantity like redundancy that decision scientists universally agree on as being a key factor in the comparison of competing alternatives," Yu said.

She suggests that uncertainty in how we value different options is behind some of our poor decisions. "If you've bought a lot of houses, you'll evaluate houses differently than if you were a first-time homebuyer," Yu said. "Or if your parents bought a house during the housing crisis, it may later affect how you buy a house."

Moreover, Yu argues, the visual and decision-making systems have different end-goals. "Vision is a sensory system whose job is to recover as much information as possible from the world," she said. "Decision-making is about trying to make a decision you'll enjoy. I think the computational goal is not just information, it's something more behaviorally relevant like total enjoyment."

For many of us, the main concern over decision-making is practical—how can we make better decisions? Glimcher said that his research has helped him develop specific strategies. "Rather than pick what I hope is the best, instead I now always start by eliminating the worst element from a choice set," he said, reducing the number of options to something manageable, like three. "I find that this really works, and it derives from our study of the math. Sometimes you learn something simple from the most complex stuff, and it really can improve your decision-making."

INFANT BRAINS REVEAL HOW THE MIND GETS BUILT

Courtney Humphries

R ebecca Saxe's first son, Arthur, was just a month old when he first entered the bore of an MRI machine to have his brain scanned. Saxe, a cognitive scientist at the Massachusetts Institute of Technology, went head-first with him: lying uncomfortably on her stomach, her face near his diaper, she stroked and soothed him as the three-tesla magnet whirred around them. Arthur, unfazed, promptly fell asleep.

All parents wonder what's going on inside their baby's mind; few have the means to find out. When Saxe got pregnant, she'd already been working with colleagues for years to devise a setup to image brain activity in babies. But her due date in September 2013 put an impetus on getting everything ready.

Over the past couple of decades, researchers like Saxe have used functional MRI to study brain activity in adults and children. But fMRI, like a 19th-century daguerreotype, requires subjects to lie perfectly still lest the image become hopelessly blurred. Babies are jittering bundles of motion when not asleep, and they can't be cajoled or bribed into stillness. The few fMRI studies done on babies to date mostly focused on playing sounds to them while they slept.

But Saxe wanted to understand how babies see the world when they're awake; she wanted to image Arthur's brain as he looked at video clips, the kind of thing that adult research subjects do easily. It was a way of approaching an even bigger question: Do babies' brains work like miniature versions of adult brains, or are they completely different? "I had this fundamental question about how brains develop, and I had a baby with a developing brain," she said. "Two of the things that were most important to me in life temporarily had this very intense convergence inside an MRI machine."

Saxe spent her maternity leave hanging out with Arthur in the machine. "Some of those days, he didn't feel like it, or he fell asleep, or he was fussy, or he pooped," she said. "Getting good data from a baby's brain is a very rare occurrence." Between sessions, Saxe and her colleagues pored over their data, tweaking their experiments, searching for a pattern in Arthur's brain activity. When they got their first usable result when he was 4 months old, she said, "I was through the roof."

A paper published in January 2017 in *Nature Communications* is the culmination of more than two years of work to image brain activity in Arthur and eight other babies.[1] In it, her team finds some surprising similarities in how the babies' and adults' brains respond to visual information, as well as some intriguing differences. The study is a first step in what Saxe hopes will become a broader effort to understand the earliest beginnings of the mind.

ORGANIZED FROM BIRTH?

Functional MRI is perhaps the most powerful tool scientists have to study brain activity short of opening up the skull. It relies on changes in blood flow in areas of the brain that are more active than others, which creates a detectable signal in an MRI machine. The technique has generated some criticism, as it's an indirect measure of brain activity, and the simple, striking images it produces rely on behind-the-scenes statistical manipulation. Nevertheless, fMRI has opened up entirely new avenues of research by giving scientists what Saxe calls "a moving map of the human brain." It has revealed, in incredible detail, how different parts of the brain choreograph their activity depending on what a person is doing, perceiving or thinking.

Some areas of the cortex also seem to be purpose-specific. Nancy Kanwisher, a neuroscientist at MIT and Saxe's former adviser, is known for discovering an area called the fusiform face area, which responds to images of faces more than any other visual input. Her lab also led the discovery of the parahippocampal place area, which preferentially responds to scenes depicting places. When she was a graduate student in Kanwisher's lab, Saxe discovered an area of the brain that was devoted to "theory of mind"— thinking about what other people are thinking. Since then, research by several labs has identified regions of the brain involved in social judgments and decision-making.

Saxe, who speaks rapidly and radiates an intellectual intensity, gets most animated by philosophical and deeply fundamental questions about the brain. For her, the next obvious question is: How did the brain's organization come about? "Seeing in adults these incredibly rich and abstract functions of your brain—morality, theory of mind—it just raises the question, how does that get there?" she said.

Have our brains evolved to have special areas devoted to the things most important for our survival? Or, she said, "is it that we were born with an amazing multipurpose learning machinery that could learn whatever organization of the world it was given to discover?" Do we enter the world with an innate blueprint for devoting parts of our brains to faces, for instance, or do we develop a specialized face area after months or years of seeing so many people around us? "The basic structural organization of the human brain could be similar from person to person because the world is similar from person to person," she said. Or its outlines could be there from birth.

A PLACE FOR FACES

Riley LeBlanc spits out her pacifier and starts to cry. A 5-month-old with a mop of curly brown hair, she's fussing in her swaddle as Heather Kosakowski, Saxe's lab manager, stands by the hulking MRI machine, housed in the bottom floor of MIT's Brain and Cognitive Sciences building, and bounces Riley up and down. Lori Fauci, Riley's mother, seated on the scanning bed, pulls another pacifier from her back pocket to offer her child.

Everything here is designed to soothe Riley. The room is softly lit, and speakers play tinkling, toy-piano versions of pop songs as lullabies (currently: Guns N' Roses' "Sweet Child o' Mine"). On the scanning bed lies a specially designed radio-frequency coil—an angled lounger and baby-sized helmet—to act as an antenna for radio signals during scans. The MRI machine is programmed with special protocols that generate less noise than usual, to avoid harming the babies' delicate hearing.

It takes a few false starts before Riley is willing to lie in the coil without fussing. Her mother positions herself on her stomach with her face and hands near Riley to soothe her. Kosakowski slides mother and child into the scanner and moves to a windowed anteroom, while Lyneé Herrera, another lab member, stays in the MRI room and gives hand signals to Kosakowski to let her know when Riley's eyes are open and watching the mirror above her head, which reflects images projected from the back of the machine.

The team's goal is to collect about 10 minutes of data from each baby while she's motionlessly watching the videos. To achieve that, the researchers often need to average together data from multiple two-hour sessions. "The more times a baby comes, the more likely we are to get that full 10 minutes," Kosakowski said. This is Riley's eighth visit.

When Herrera signals that Riley has stopped napping, Kosakowski initiates the scanner and cues a series of video clips, as babies are more likely to look at moving images than still ones. After a while, Herrera closes her hand, signaling that Riley's eyes are closed again. "Sometimes I think babies must be getting their best naps here," Kosakowski said with a laugh.

Studying infants has always required creative techniques. "It's been an interesting problem," said Charles Nelson, a cognitive neuroscientist at Harvard Medical School and Boston Children's Hospital who studies child development, "because you're dealing with a nonverbal, rhetorically limited, attentionally limited organism, trying to figure out what's going on inside their head." Similar techniques are often used to study babies and nonhuman primates, or children with disabilities who are not verbal. "We have a class of covert measures that allows us to peek inside the monkey, the baby, the child with a disorder," Nelson said.

The simplest is watching their behavior and noting where they look, either by observing them or using eye-tracking technologies. Another is to measure brain activity. Electroencephalography (EEG), for instance, simply requires attaching an adorable skullcap of electrodes and wires to a baby's head to detect fluctuating brain waves.[2] And a newer technique called near infrared spectroscopy (NIRS) sends light through babies' thin, soft skulls to detect changes in blood flow in the brain.

Both methods reveal how brain activity changes moment to moment, but NIRS only reaches the outer layers of the brain, and EEG can't show exactly which brain areas are active. "To study the detailed spatial organization, and to get to deeper brain regions, you have to go to fMRI," said Ben Deen, first author of the *Nature Communications* study, who's now a researcher at Rockefeller University.

Using other methods, researchers have found hints that babies respond differently to visual inputs of different categories, particularly faces. Faces "are a very salient part of environment," said Michelle de Haan, a developmental neuroscientist at University College London. In the first few weeks of life, an infant's eyes focus best on objects around the distance of a nursing mother's face. Some researchers believe babies may have an innate mechanism, deep in the brain, which directs their eyes to look at faces.

There's evidence that young infants will look longer at faces than other things. A baby's response to faces also becomes more specialized over time and with experience. For instance, adults have a harder time discriminating between two faces when they're upside down, but babies under 4 months of age don't have this bias—they can discriminate between two upside-down faces as easily as two right-side-up ones. After about 4 months, though, they acquire a bias for right-side-up faces. Around 6 months of age, infants who see faces produce an EEG signature of activity that is similar to that of adults who see faces.

But while this research suggests that babies might have some specialization in their brain for certain categories like faces, Deen said, "we knew very little about the detail of where those signals are coming from."

For their paper, Saxe and her colleagues obtained data from nine of the 17 babies they scanned. Though the lab is increasingly relying on outside families recruited into studies, it helped that they had a spate of "lab babies" to start with, including Arthur; Saxe's second son, Percy; her sister's son; and a postdoc's son. They presented babies with movies of faces, natural scenes, human bodies and objects—toys, in this case—as well as scrambled scenes, in which parts of the image are jumbled. Saxe said they focused on faces versus scenes because the two stimuli create a sharp difference in adult brains, evoking activity in very different regions.

Surprisingly, they found a similar pattern in babies. "Every region that we knew about in adults [with] a preference for faces or scenes has that same preference in babies 4 to 6 months old," Saxe said. That shows that the cortex "is already starting to have a bias in its function," she said, rather than being totally undifferentiated.

Are babies born with this ability? "We can't strictly say that anything is innate," Deen said. "We can say it develops very early." And Saxe points out that the responses extended beyond the visual cortex (the structures of the brain responsible for directly processing visual inputs). The researchers also found differences in the frontal cortex, an area of the brain involved in emotions, values and self-representation. "To see frontal cortex engagement in a baby is really exciting," she said. "It's thought to be one of the last spots to fully develop."

However, while Saxe's team found that similar areas of the brain were active in babies and adults, they did not find evidence that infants have areas specialized for one particular input, like faces or scenes, over all others. Nelson, who was not involved in the study, said it suggests that infant brains are "more multipurpose." He added: "That points out a fundamental difference in the infant brain versus the adult brain."

THE FLEXIBLE BRAIN

It's surprising that babies' brains behave like adults' brains at all considering how different they look. On a computer screen outside the MRI room at MIT, I can see anatomical images of Riley's brain that were taken while she napped. Compared to MRI scans of adult brains, in which different brain structures are clearly visible, Riley's brain seems creepily dark.

"It looks like this is just a really poor image, doesn't it?" Kosakowski said. She explains that babies at this stage have not yet fully developed the fatty insulation around nerve fibers, called myelin, that makes up the brain's white matter. The corpus callosum, a yoke of nerve fibers connecting the two hemispheres of the brain, is only dimly visible.

At this age, a baby's brain is expanding—the cerebral cortex swells by 88 percent in the first year of life. Its cells are also reorganizing themselves and rapidly forming new connections to one another, many of which will get winnowed back throughout childhood and adolescence. At this stage, the brain is astonishingly flexible: when babies have strokes or seizures that require having an entire hemisphere of the brain surgically removed, they recover remarkably well. But there are also limits to this flexibility; babies who experience deprivation or abuse may have lifelong learning deficits.

Studying how healthy human brains develop can help scientists understand why this process sometimes goes awry. It's known, for instance, that many children and adults with autism have difficulties with social cognition tasks, such as interpreting faces.[3] Are these differences present at the earliest stages of the brain's development, or do they emerge out of a child's experience, driven by a lack of attention to faces and social cues?

We're only beginning to understand how babies' brains are organized; it will require many more hours collecting data from a larger number of babies to have a fuller picture of how their brains work. But Saxe and her colleagues have shown that such a study can be done, which opens up new areas of investigation. "It is possible to get good fMRI data in awake babies—if you are extremely patient," Saxe said. "Now let's try to figure out what we can learn from it."

VI HOW DO MACHINES LEARN?

IS ALPHAGO REALLY SUCH A BIG DEAL?

Michael Nielsen

I n 1997, IBM's Deep Blue system defeated the world chess champion, Garry Kasparov. At the time, the victory was widely described as a milestone in artificial intelligence. But Deep Blue's technology turned out to be useful for chess and not much else. Computer science did not undergo a revolution.

Will AlphaGo, the Go-playing system that recently defeated one of the strongest Go players in history, be any different?

I believe the answer is yes, but not for the reasons you may have heard. Many articles proffer expert testimony that Go is harder than chess, making this victory more impressive. Or they say that we didn't expect computers to win at Go for another 10 years, so this is a bigger breakthrough. Some articles offer the (correct!) observation that there are more potential positions in Go than in chess, but they don't explain why this should cause more difficulty for computers than for humans.

In other words, these arguments don't address the core question: Will the technical advances that led to AlphaGo's success have broader implications? To answer this question, we must first understand the ways in which the advances that led to AlphaGo are qualitatively different and more important than those that led to Deep Blue.

In chess, beginning players are taught a notion of a chess piece's value. In one system, a knight or bishop is worth three pawns. A rook, which has greater range of movement, is worth five pawns. And the queen, which has the greatest range of all, is worth nine pawns. A king has infinite value, since losing it means losing the game.

You can use these values to assess potential moves. Give up a bishop to take your opponent's rook? That's usually a good idea. Give up a knight and a bishop in exchange for a rook? Not such a good idea.

The notion of value is crucial in computer chess. Most computer chess programs search through millions or billions of combinations of moves and countermoves. The goal is for the program to find a sequence of moves that maximizes the final value of the program's board position, no matter what sequence of moves is played by the opponent.

Early chess programs evaluated board positions using simple notions like "one bishop equals three pawns." But later programs used more detailed chess knowledge. Deep Blue, for example, combined more than 8,000 different factors in the function it used to evaluate board positions. Deep Blue didn't just say that one rook equals five pawns. If a pawn of the same color is ahead of the rook, the pawn will restrict the rook's range of movement, thus making the rook a little less valuable. If, however, the pawn is "levered," meaning that it can move out of the rook's way by capturing an enemy pawn, Deep Blue considers the pawn semitransparent and doesn't reduce the rook's value as much.

Ideas like this depend on detailed knowledge of chess and were crucial to Deep Blue's success. According to the technical paper written by the Deep Blue team, this notion of a semitransparent levered pawn was crucial to Deep Blue's play in the second game against Kasparov.

Ultimately, the Deep Blue developers used two main ideas. The first was to build a function that incorporated lots of detailed chess knowledge to evaluate any given board position. The second was to use immense computing power to evaluate lots of possible positions, picking out the move that would force the best possible final board position.

What happens if you apply this strategy to Go?

It turns out that you will run into a difficult problem when you try. The problem lies in figuring out how to evaluate board positions. Top Go players use a lot of intuition in judging how good a particular board position is. They will, for instance, make vague-sounding statements about a board position having "good shape." And it's not immediately clear how to express this intuition in simple, well-defined systems like the valuation of chess pieces.

Now you might think it's just a question of working hard and coming up with a good way of evaluating board positions. Unfortunately, even after decades of attempts to do this using conventional approaches, there was still no obvious way to apply the search strategy that was so successful for chess, and Go programs remained disappointing. This began to change in 2006, with the introduction of so-called Monte Carlo tree search algorithms, which tried a new approach to evaluation based on a clever way of randomly simulating games. But Go programs still fell far short of human players in ability. It seemed as though a strong intuitive sense of board position was essential to success.

What's new and important about AlphaGo is that its developers have figured out a way of bottling something very like that intuitive sense.

To explain how it works, let me describe the AlphaGo system, as outlined in the paper the AlphaGo team published in January 2016.[1] (The details of the system were somewhat improved for AlphaGo's match against Lee Sedol, but the broad governing principles remain the same.)

To begin, AlphaGo took 150,000 games played by good human players and used an artificial neural network to find patterns in those games. In particular, it learned to predict with high probability what move a human player would take in any given position. AlphaGo's designers then improved the neural network by repeatedly playing it against earlier versions of itself, adjusting the network so it gradually improved its chance of winning.

How does this neural network—known as the policy network—learn to predict good moves?

Broadly speaking, a neural network is a very complicated mathematical model, with millions of parameters that can be adjusted to change the model's behavior. When I say the network "learned," what I mean is that the computer kept making tiny adjustments to the parameters in the model, trying to find a way to make corresponding tiny improvements in its play. In the first stage of learning, the network tried to increase the probability of making the same move as the human players. In the second stage, it tried to increase the probability of winning a game in self-play. This sounds like a crazy strategy—repeatedly making tiny tweaks to some enormously complicated function—but if you do this for long enough, with enough computing power, the network gets pretty good. And here's the strange thing: It gets good for reasons no one really understands, since the improvements are a consequence of billions of tiny adjustments made automatically.

After these two training stages, the policy network could play a decent game of Go, at the same level as a human amateur. But it was still a long way from professional quality. In a sense, it was a way of playing Go without searching through future lines of play and estimating the value of the resulting board positions. To improve beyond the amateur level, AlphaGo needed a way of estimating the value of those positions.

To get over this hurdle, the developers' core idea was for AlphaGo to play the policy network against itself, to get an estimate of how likely a given board position was to be a winning one. That probability of a win provided a rough valuation of the position. (In practice, AlphaGo used a slightly more complex variation of this idea.) Then, AlphaGo combined this approach to valuation with a search through many possible lines of play, biasing its search toward lines of play the policy network thought were likely. It then picked the move that forced the highest effective board valuation.

We can see from this that AlphaGo didn't start out with a valuation system based on lots of detailed knowledge of Go, the way Deep Blue did for chess. Instead, by analyzing thousands of prior games and engaging in a lot of self-play, AlphaGo created a policy network through billions of tiny adjustments, each intended to make just a tiny incremental improvement. That, in turn, helped AlphaGo build a valuation system that captures something very similar to a good Go player's intuition about the value of different board positions.

In this way, AlphaGo is much more radical than Deep Blue. Since the earliest days of computing, computers have been used to search out ways of optimizing known functions. Deep Blue's approach was just that: a search aimed at optimizing a function whose form, while complex, mostly expressed existing chess knowledge. It was clever about how it did this search, but it wasn't that different from many programs written in the 1960s.

AlphaGo also uses the search-and-optimization idea, although it is somewhat cleverer about how it does the search. But what is new and unusual is the prior stage, in which it uses a neural network to learn a function that helps capture some sense of good board position. It was by combining those two stages that AlphaGo became able to play at such a high level.

This ability to replicate intuitive pattern recognition is a big deal. It's also part of a broader trend. In an earlier paper, the same organization that built AlphaGo—Google DeepMind—built a neural network that learned to play 49 classic Atari 2600 video games, in many cases reaching a level that human experts couldn't match.[2] The conservative approach to solving this problem with a computer would be in the style of Deep Blue: A human programmer would analyze each game and figure out detailed control strategies for playing it.

By contrast, DeepMind's neural network simply explored lots of ways of playing. Initially, it was terrible, flailing around wildly, rather like a human newcomer. But occasionally the network would accidentally do clever things. It learned to recognize good patterns of play—in other words, patterns leading to higher scores—in a manner not unlike the way AlphaGo learned good board position. And when that happened, the network would reinforce the behavior, gradually improving its ability to play.

This ability of neural networks to bottle intuition and pattern recognition is being used in other contexts. In 2015, Leon Gatys, Alexander Ecker and Matthias Bethge posted a paper to the scientific preprint site arxiv.org describing a way for a neural network to learn artistic styles and then to apply those styles to other images.[3] The idea was very simple: The network was exposed to a very large number of images and acquired an ability to recognize images with similar styles. It could then apply that style information to new images.

It's not great art, but it's still a remarkable example of using a neural network to capture an intuition and apply it elsewhere.

Over the past few years, neural networks have been used to capture intuition and recognize patterns across many domains. Many of the projects employing these networks have been visual in nature, involving tasks such as recognizing artistic style or developing good video-game strategy. But there are also striking examples of networks simulating intuition in very different domains, including audio and natural language.

Because of this versatility, I see AlphaGo not as a revolutionary breakthrough in itself, but rather as the leading edge of an extremely important development: the ability to build systems that can capture intuition and learn to recognize patterns. Computer scientists have attempted to do this for decades, without making much progress. But now, the success of neural networks has the potential to greatly expand the range of problems we can use computers to attack.

It's tempting at this point to cheer wildly, and to declare that general artificial intelligence must be just a few years away. After all, suppose you divide up ways of thinking into logical thought of the type we already know computers are good at, and "intuition." If we view AlphaGo and similar systems as proving that computers can now simulate intuition, it seems as though all bases are covered: Computers can now perform both logic and intuition. Surely general artificial intelligence must be just around the corner!

But there's a rhetorical fallacy here: We've lumped together many different mental activities as "intuition." Just because neural networks can do a good job of capturing some specific types of intuition, that doesn't mean they can do as good a job with other types. Maybe neural networks will be no good at all at some tasks we currently think of as requiring intuition.

In actual fact, our existing understanding of neural networks is very poor in important ways. For example, a 2014 paper described certain "adversarial examples" which can be used to fool neural networks.[4] The authors began their work with a neural network that was extremely good at recognizing images. It seemed like a classic triumph of using neural networks to capture pattern-recognition ability. But what they showed is that it's possible to fool the network by changing images in tiny ways.

Another limitation of existing systems is that they often require many human examples to learn from. For instance, AlphaGo learned from 150,000 human games. That's a lot of games! By contrast, human beings can learn a great deal from far fewer games. Similarly, networks that recognize and manipulate images are typically trained on millions of example images, each annotated with information about the image type. And so an important challenge is to make the systems better at learning from smaller human-supplied data sets, and with less ancillary information.

With that said, systems like AlphaGo are genuinely exciting. We have learned to use computer systems to reproduce at least some forms of human intuition. Now we've got so many wonderful challenges ahead: to expand the range of intuition types we can represent, to make the systems stable, to understand why and how they work and to learn better ways to combine them with the existing strengths of computer systems. Might we soon learn to capture some of the intuitive judgment that goes into writing mathematical proofs, or into writing stories or good explanations? It's a tremendously promising time for artificial intelligence.

NEW THEORY CRACKS OPEN THE BLACK BOX OF DEEP LEARNING

Natalie Wolchover

E ven as machines known as "deep neural networks" have learned to converse, drive cars, beat video games and Go champions, dream, paint pictures and help make scientific discoveries, they have also confounded their human creators, who never expected so-called "deep-learning" algorithms to work so well. No underlying principle has guided the design of these learning systems, other than vague inspiration drawn from the architecture of the brain (and no one really understands how that operates either).

Like a brain, a deep neural network has layers of neurons—artificial ones that are figments of computer memory. When a neuron fires, it sends signals to connected neurons in the layer above. During deep learning, connections in the network are strengthened or weakened as needed to make the system better at sending signals from input data—the pixels of a photo of a dog, for instance—up through the layers to neurons associated with the right high-level concepts, such as "dog." After a deep neural network has "learned" from thousands of sample dog photos, it can identify dogs in new photos as accurately as people can. The magic leap from special cases to general concepts during learning gives deep neural networks their power, just as it underlies human reasoning, creativity and the other faculties collectively termed "intelligence." Experts wonder what it is about deep learning that enables generalization—and to what extent brains apprehend reality in the same way.

In August 2017, a YouTube video of a conference talk in Berlin, shared widely among artificial-intelligence researchers, offered a possible answer. In the talk, Naftali Tishby, a computer scientist and neuroscientist from the Hebrew University of Jerusalem, presented evidence in support of a new theory explaining how deep learning works. Tishby argues that deep neural networks learn according to a procedure called the "information bottleneck," which he and two collaborators first described in purely theoretical terms in 1999.[1] The idea is that a network rids noisy input data of extraneous details as if by squeezing the information through a bottleneck, retaining only the features most relevant to general concepts. Striking computer

Learning from Experience

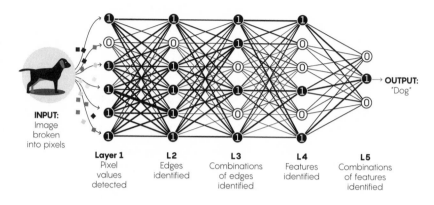

FIGURE 6.1

Deep neural networks learn by adjusting the strengths of their connections to better convey input signals through multiple layers to neurons associated with the right general concepts. When data is fed into a network, each artificial neuron that fires (labeled "1") transmits signals to certain neurons in the next layer, which are likely to fire if multiple signals are received. The process filters out noise and retains only the most relevant features.

experiments by Tishby and his student Ravid Shwartz-Ziv reveal how this squeezing procedure happens during deep learning, at least in the cases they studied.[2]

Tishby's findings have the AI community buzzing. "I believe that the information bottleneck idea could be very important in future deep neural network research," said Alex Alemi of Google Research, who has already developed new approximation methods for applying an information bottleneck analysis to large deep neural networks.[3] The bottleneck could serve "not only as a theoretical tool for understanding why our neural networks work as well as they do currently, but also as a tool for constructing new objectives and architectures of networks," Alemi said.

Some researchers remain skeptical that the theory fully accounts for the success of deep learning, but Kyle Cranmer, a particle physicist at New York University who uses machine learning to analyze particle collisions at the Large Hadron Collider, said that as a general principle of learning, it "somehow smells right."

Geoffrey Hinton, a pioneer of deep learning who works at Google and the University of Toronto, emailed Tishby after watching his Berlin talk. "It's extremely interesting," Hinton wrote. "I have to listen to it another 10,000 times to really understand it, but it's very rare nowadays to hear a talk with a really original idea in it that may be the answer to a really major puzzle."

According to Tishby, who views the information bottleneck as a fundamental principle behind learning, whether you're an algorithm, a housefly, a conscious being or a physics calculation of emergent behavior, that long-awaited answer "is that the most important part of learning is actually forgetting."

THE BOTTLENECK

Tishby began contemplating the information bottleneck around the time that other researchers were first mulling over deep neural networks, though neither concept had been named yet. It was the 1980s, and Tishby was thinking about how good humans are at speech recognition—a major challenge for AI at the time. Tishby realized that the crux of the issue was the question of relevance: What are the most relevant features of a spoken word, and how do we tease these out from the variables that accompany them, such as accents, mumbling and intonation? In general, when we face the sea of data that is reality, which signals do we keep?

"This notion of relevant information was mentioned many times in history but never formulated correctly," Tishby said in a 2017 interview. "For many years people thought information theory wasn't the right way to think about relevance, starting with misconceptions that go all the way to Shannon himself."

Claude Shannon, the founder of information theory, in a sense liberated the study of information starting in the 1940s by allowing it to be considered in the abstract—as 1s and 0s with purely mathematical meaning. Shannon took the view that, as Tishby put it, "information is not about semantics." But, Tishby argued, this isn't true. Using information theory, he realized, "you can define 'relevant' in a precise sense."

Imagine X is a complex data set, like the pixels of a dog photo, and Y is a simpler variable represented by those data, like the word "dog." You can capture all the "relevant" information in X about Y by compressing X as much as you can without losing the ability to predict Y. In their 1999 paper, Tishby and co-authors Fernando Pereira, now at Google, and William Bialek, now at Princeton University, formulated this as a mathematical optimization problem. It was a fundamental idea with no killer application.

"I've been thinking along these lines in various contexts for 30 years," Tishby said. "My only luck was that deep neural networks became so important."

EYEBALLS ON FACES ON PEOPLE ON SCENES

Though the concept behind deep neural networks had been kicked around for decades, their performance in tasks like speech and image recognition only took off in the early 2010s, due to improved training regimens and more powerful computer processors. Tishby recognized their potential connection to the information bottleneck principle in 2014 after reading a surprising paper by the physicists David Schwab and Pankaj Mehta.[4]

The duo discovered that a deep-learning algorithm invented by Hinton called the "deep belief net" works, in a particular case, exactly like renormalization, a technique used in physics to zoom out on a physical system by coarse-graining over its details and calculating its overall state. When Schwab and Mehta applied the deep belief net to a model of a magnet at its "critical point," where the system is fractal, or self-similar at every scale, they found that the network automatically used the renormalization-like procedure to discover the model's state. It was a stunning indication that, as the biophysicist Ilya Nemenman said at the time, "extracting relevant features in the context of statistical physics and extracting relevant features in the context of deep learning are not just similar words, they are one and the same."

The only problem is that, in general, the real world isn't fractal. "The natural world is not ears on ears on ears on ears; it's eyeballs on faces on people on scenes," Cranmer said. "So I wouldn't say [the renormalization procedure] is why deep learning on natural images is working so well." But Tishby, who at the time was undergoing chemotherapy for pancreatic cancer, realized that both deep learning and the coarse-graining procedure could be encompassed by a broader idea. "Thinking about science and about the role of my old ideas was an important part of my healing and recovery," he said.

In 2015, he and his student Noga Zaslavsky hypothesized that deep learning is an information bottleneck procedure that compresses noisy data as much as possible while preserving information about what the data represent.[5] Tishby and Shwartz-Ziv's newer experiments with deep neural networks reveal how the bottleneck procedure actually plays out. In one case, the researchers used small networks that could be trained to label input data with a 1 or 0 (think "dog" or "no dog") and gave their 282 neural connections random initial strengths. They then tracked what happened as the networks engaged in deep learning with 3,000 sample input data sets.

The basic algorithm used in the majority of deep-learning procedures to tweak neural connections in response to data is called "stochastic gradient descent": Each time the training data are fed into the network, a cascade of firing activity sweeps upward through the layers of artificial neurons. When the signal reaches the top layer, the final firing pattern can be compared to the correct label for the image—1 or 0, "dog" or "no dog." Any differences between this firing pattern and the correct pattern are "back-propagated" down the layers, meaning that, like a teacher correcting an exam, the algorithm strengthens or weakens each connection to make the network layer better at producing the correct output signal. Over the course of training, common patterns in the training data become reflected in the strengths of the connections, and the network becomes expert at correctly labeling the data, such as by recognizing a dog, a word or a 1.

In their experiments, Tishby and Shwartz-Ziv tracked how much information each layer of a deep neural network retained about the input data and how much information each one retained about the output label. The scientists found that, layer by layer, the networks converged to the information bottleneck theoretical bound: a theoretical limit derived in Tishby, Pereira and Bialek's original paper that represents the absolute best the system can do at extracting relevant information. At the bound, the network has compressed the input as much as possible without sacrificing the ability to accurately predict its label.

Tishby and Shwartz-Ziv also made the intriguing discovery that deep learning proceeds in two phases: a short "fitting" phase, during which the network learns to label its training data, and a much longer "compression" phase, during which it becomes good at generalization, as measured by its performance at labeling new test data.

As a deep neural network tweaks its connections by stochastic gradient descent, at first the number of bits it stores about the input data stays roughly constant or increases slightly, as connections adjust to encode patterns in the input and the network gets good at fitting labels to it. Some experts have compared this phase to memorization.

Then learning switches to the compression phase. The network starts to shed information about the input data, keeping track of only the strongest features—those correlations that are most relevant to the output label. This happens because, in each iteration of stochastic gradient descent, more or less accidental correlations in the training data tell the network to do different things, dialing the strengths of its neural connections up and down in a random walk. This randomization is effectively the same as compressing the system's representation of the input data. As an example, some photos of dogs might have houses in the background, while others don't. As a network cycles through these training photos, it might "forget" the correlation between houses and dogs in some photos as other photos counteract it. It's this forgetting of specifics, Tishby and Shwartz-Ziv argue, that enables the system to form general concepts. Indeed, their experiments revealed that deep neural networks ramp up their generalization performance during the compression phase, becoming better at labeling test data. (A deep neural network trained to recognize dogs in photos might be tested on new photos that may or may not include dogs, for instance.)

It remains to be seen whether the information bottleneck governs all deep-learning regimes, or whether there are other routes to generalization besides compression. Some AI experts see Tishby's idea as one of many important theoretical insights about deep learning to have emerged recently. Andrew Saxe, an AI researcher and theoretical neuroscientist at Harvard University, noted that certain very large deep neural networks don't seem to need a drawn-out compression phase in order to generalize well. Instead, researchers program in something called early stopping, which cuts training short to prevent the network from encoding too many correlations in the first place.

Tishby argues that the network models analyzed by Saxe and his colleagues differ from standard deep neural network architectures, but that nonetheless, the information bottleneck theoretical bound defines these networks' generalization performance better than other methods. Questions about whether the bottleneck holds up for larger neural networks are partly addressed by Tishby and Shwartz-Ziv's more recent experiments, not included in their preliminary paper, in which they train much larger, 330,000-connection-deep neural networks to recognize handwritten digits in the 60,000-image Modified National Institute of Standards and Technology database, a well-known benchmark for gauging the performance of deep-learning algorithms. The scientists saw the same convergence of the networks to the information bottleneck theoretical bound; they also observed the two distinct phases of deep learning, separated by an even sharper transition than in the smaller networks. "I'm completely convinced now that this is a general phenomenon," Tishby said.

HUMANS AND MACHINES

The mystery of how brains sift signals from our senses and elevate them to the level of our conscious awareness drove much of the early interest in deep neural networks among AI pioneers, who hoped to reverse-engineer the brain's learning rules. AI practitioners have since largely abandoned that path in the mad dash for technological progress, instead slapping on bells and whistles that boost performance with little regard for biological plausibility. Still, as their thinking machines achieve ever greater feats—even stoking fears that AI could someday pose an existential threat—many researchers hope these explorations will uncover general insights about learning and intelligence.

Brenden Lake, an assistant professor of psychology and data science at New York University who studies similarities and differences in how humans and machines learn, said that Tishby's findings represent "an important step towards opening the black box of neural networks," but he stressed that the brain represents a much bigger, blacker black box. Our adult brains, which boast several hundred trillion connections between 86 billion neurons, in all likelihood employ a bag of tricks to enhance generalization, going beyond the basic image- and sound-recognition learning procedures that occur during infancy and that may in many ways resemble deep learning.

For instance, Lake said the fitting and compression phases that Tishby identified don't seem to have analogues in the way children learn handwritten characters, which he studies. Children don't need to see thousands of examples of a character and compress their mental representation over an extended period of time before they're able to recognize other instances of that letter and write it themselves. In fact, they can learn from a single example. Lake and his colleagues' models suggest the brain may deconstruct the new letter into a series of strokes—previously existing mental constructs—allowing the conception of the letter to be tacked onto an edifice of prior knowledge.[6] "Rather than thinking of an image of a letter as a pattern of pixels and learning the concept as mapping those features" as in standard machine-learning algorithms, Lake explained, "instead I aim to build a simple causal model of the letter," a shorter path to generalization.

Such brainy ideas might hold lessons for the AI community, furthering the back-and-forth between the two fields. Tishby believes his information bottleneck theory will ultimately prove useful in both disciplines, even if it takes a more general form in human learning than in AI. One immediate insight that can be gleaned from the theory is a better understanding of which kinds of problems can be solved by real and artificial neural networks.

"It gives a complete characterization of the problems that can be learned," Tishby said. These are "problems where I can wipe out noise in the input without hurting my ability to classify. This is natural vision problems, speech recognition. These are also precisely the problems our brain can cope with."

Meanwhile, both real and artificial neural networks stumble on problems in which every detail matters and minute differences can throw off the whole result. Most people can't quickly multiply two large numbers in their heads, for instance. "We have a long class of problems like this, logical problems that are very sensitive to changes in one variable," Tishby said. "Classifiability, discrete problems, cryptographic problems. I don't think deep learning will ever help me break cryptographic codes."

Generalizing—traversing the information bottleneck, perhaps—means leaving some details behind. This isn't so good for doing algebra on the fly, but that's not a brain's main business. We're looking for familiar faces in the crowd, order in chaos, salient signals in a noisy world.

A BRAIN BUILT FROM ATOMIC SWITCHES CAN LEARN

Andreas von Bubnoff

B rains, beyond their signature achievements in thinking and problem solving, are paragons of energy efficiency. The human brain's power consumption resembles that of a 20-watt incandescent lightbulb. In contrast, one of the world's largest and fastest supercomputers, the K computer in Kobe, Japan, consumes as much as 9.89 megawatts of energy—an amount roughly equivalent to the power usage of 10,000 households. Yet in 2013, even with that much power, it took the machine 40 minutes to simulate just a single second's worth of 1 percent of human brain activity.

Now engineering researchers at the California NanoSystems Institute at the University of California, Los Angeles, are hoping to match some of the brain's computational and energy efficiency with systems that mirror the brain's structure. They are building a device, perhaps the first one, that is "inspired by the brain to generate the properties that enable the brain to do what it does," according to Adam Stieg, a research scientist and associate director of the institute, who leads the project with Jim Gimzewski, a professor of chemistry at UCLA.[1]

The device is a far cry from conventional computers, which are based on minute wires imprinted on silicon chips in highly ordered patterns. The current pilot version is a 2-millimeter-by-2-millimeter mesh of silver nanowires connected by artificial synapses. Unlike silicon circuitry, with its geometric precision, this device is messy, like "a highly interconnected plate of noodles," Stieg said. And instead of being designed, the fine structure of the UCLA device essentially organized itself out of random chemical and electrical processes.

Yet in its complexity, this silver mesh network resembles the brain. The mesh boasts 1 billion artificial synapses per square centimeter, which is within a couple of orders of magnitude of the real thing. The network's electrical activity also displays a property unique to complex systems like the brain: "criticality," a state between order and chaos indicative of maximum efficiency.

Moreover, preliminary experiments suggest that this neuromorphic (brainlike) silver wire mesh has great functional potential. It can already perform simple learning and logic operations. It can clean the unwanted noise from received signals, a capability that's important for voice recognition and similar tasks that challenge conventional computers. And its existence proves the principle that it might be possible one day to build devices that can compute with an energy efficiency close to that of the brain.

These advantages look especially appealing as the limits of miniaturization and efficiency for silicon microprocessors now loom. "Moore's law is dead, transistors are no longer getting smaller, and [people] are going, 'Oh, my God, what do we do now?'" said Alex Nugent, CEO of the Santa Fe-based neuromorphic computing company Knowm, who was not involved in the UCLA project. "I'm very excited about the idea, the direction of their work," Nugent said. "Traditional computing platforms are a billion times less efficient."

SWITCHES THAT ACT LIKE SYNAPSES

Energy efficiency wasn't Gimzewski's motivation when he started the silver wire project 10 years ago. Rather, it was boredom. After using scanning tunneling microscopes to look at electronics at the atomic scale for 20 years, he said, "I was tired of perfection and precise control [and] got a little bored with reductionism."

In 2007, he accepted an invitation to study single atomic switches developed by a group that Masakazu Aono led at the International Center for Materials Nanoarchitectonics in Tsukuba, Japan. The switches contain the same ingredient that turns a silver spoon black when it touches an egg: silver sulfide, sandwiched between solid metallic silver.

Applying voltage to the devices pushes positively charged silver ions out of the silver sulfide and toward the silver cathode layer, where they are reduced to metallic silver. Atom-wide filaments of silver grow, eventually closing the gap between the metallic silver sides. As a result, the switch is on and current can flow. Reversing the current flow has the opposite effect: The silver bridges shrink, and the switch turns off.

Soon after developing the switch, however, Aono's group started to see irregular behavior. The more often the switch was used, the more easily it would turn on. If it went unused for a while, it would slowly turn off by itself. In effect, the switch remembered its history. Aono and his colleagues also found that the switches seemed to interact with each other, such that turning on one switch would sometimes inhibit or turn off others nearby.

Most of Aono's group wanted to engineer these odd properties out of the switches. But Gimzewski and Stieg (who had just finished his doctorate in Gimzewski's group) were reminded of synapses, the switches between nerve cells in the human brain, which also change their responses with experience and interact with each other. During one of their many visits to Japan, they had an idea. "We thought: Why don't we try to embed them in a structure reminiscent of the cortex in a mammalian brain [and study that]?" Stieg said.

Building such an intricate structure was a challenge, but Stieg and Audrius Avizienis, who had just joined the group as a graduate student, developed a protocol to do it. By pouring silver nitrate onto tiny copper spheres, they could induce a network of microscopically thin intersecting silver wires to grow. They could then expose the mesh to sulfur gas to create a silver sulfide layer between the silver wires, as in the Aono team's original atomic switch.

SELF-ORGANIZED CRITICALITY

When Gimzewski and Stieg told others about their project, almost nobody thought it would work. Some said the device would show one type of static activity and then sit there, Stieg recalled. Others guessed the opposite: "They said the switching would cascade and the whole thing would just burn out," Gimzewski said.

But the device did not melt. Rather, as Gimzewski and Stieg observed through an infrared camera, the input current kept changing the paths it followed through the device—proof that activity in the network was not localized but rather distributed, as it is in the brain.

Then, one fall day in 2010, while Avizienis and his fellow graduate student Henry Sillin were increasing the input voltage to the device, they suddenly saw the output voltage start to fluctuate, seemingly at random, as if the mesh of wires had come alive. "We just sat and watched it, fascinated," Sillin said.

They knew they were on to something. When Avizienis analyzed several days' worth of monitoring data, he found that the network stayed at the same activity level for short periods more often than for long periods. They later found that smaller areas of activity were more common than larger ones.

"That was really jaw-dropping," Avizienis said, describing it as "the first [time] we pulled a power law out of this." Power laws describe mathematical relationships in which one variable changes as a power of the other. They apply to systems in which larger scale, longer events are much less common than smaller scale, shorter ones—but are also still far more common than one would expect from a chance distribution. Per Bak, the Danish physicist who died in 2002, first proposed power laws as hallmarks of all kinds of complex dynamical systems that can organize over large timescales and long distances.[2,3] Power-law behavior, he said, indicates that a complex system operates at a dynamical sweet spot between order and chaos, a state of "criticality" in which all parts are interacting and connected for maximum efficiency.

As Bak predicted, power-law behavior has been observed in the human brain: In 2003, Dietmar Plenz, a neuroscientist with the National Institutes of Health, observed that groups of nerve cells activated others, which in turn activated others, often forming systemwide activation cascades. Plenz found that the sizes of these cascades fell along a power-law distribution, and that the brain was indeed operating in a way that maximized activity propagation without risking runaway activity.

The fact that the UCLA device also shows power-law behavior is a big deal, Plenz said, because it suggests that, as in the brain, a delicate balance between activation and inhibition keeps all of its parts interacting with one another. The activity doesn't overwhelm the network, but it also doesn't die out.

Gimzewski and Stieg later found an additional similarity between the silver network and the brain: Just as a sleeping human brain shows fewer short activation cascades than a brain that's awake, brief activation states in the silver network become less common at lower energy inputs. In a way, then, reducing the energy input into the device can generate a state that resembles the sleeping state of the human brain.

TRAINING AND RESERVOIR COMPUTING

But even if the silver wire network has brainlike properties, can it solve computing tasks? Preliminary experiments suggest the answer is yes, although the device is far from resembling a traditional computer.

For one thing, there is no software. Instead, the researchers exploit the fact that the network can distort an input signal in many different ways, depending on where the output is measured. This suggests possible uses for voice or image recognition, because the device should be able to clean a noisy input signal.

But it also suggests that the device could be used for a process called reservoir computing. Because one input could in principle generate many, perhaps millions, of different outputs (the "reservoir"), users can choose or combine outputs in such a way that the result is a desired computation of the inputs. For example, if you stimulate the device at two different places at the same time, chances are that one of the millions of different outputs will represent the sum of the two inputs.

The challenge is to find the right outputs and decode them and to find out how best to encode information so that the network can understand it. The way to do this is by training the device: by running a task hundreds or perhaps thousands of times, first with one type of input and then with another, and comparing which output best solves a task. "We don't program the device but we select the best way to encode the information such that the [network behaves] in an interesting and useful manner," Gimzewski said.

In work published in 2017, the researchers trained the wire network to execute simple logic operations.[4] And in other experiments, they trained the network to solve the equivalent of a simple memory task taught to lab rats called a T-maze test. In the test, a rat in a T-shaped maze is rewarded when it learns to make the correct turn in response to a light. With its own version of training, the network could make the correct response 94 percent of the time.

So far, these results aren't much more than a proof of principle, Nugent said. "A little rat making a decision in a T-maze is nowhere close to what somebody in machine learning does to evaluate their systems" on a traditional computer, he said. He doubts the device will lead to a chip that does much that's useful in the next few years.

But the potential, he emphasized, is huge. That's because the network, like the brain, doesn't separate processing and memory. Traditional computers need to shuttle information between different areas that handle the two functions. "All that extra communication adds up because it takes energy to charge wires," Nugent said. With traditional machines, he said, "literally, you could run France on the electricity that it would take to simulate a full human brain at moderate resolution." If devices like the silver wire network can eventually solve tasks as effectively as machine-learning algorithms running on traditional computers, they could do so using only one-billionth as much power. "As soon as they do that, they're going to win in power efficiency, hands down," Nugent said.

The UCLA findings also lend support to the view that under the right circumstances, intelligent systems can form by self-organization, without the need for any template or process to design them. The silver network "emerged spontaneously," said Todd Hylton, the former manager of the Defense Advanced Research Projects Agency program that supported early stages of the project. "As energy flows through [it], it's this big dance because every time one new structure forms, the energy doesn't go somewhere else.

People have built computer models of networks that achieve some critical state. But this one just sort of did it all by itself."

Gimzewski believes that the silver wire network or devices like it might be better than traditional computers at making predictions about complex processes. Traditional computers model the world with equations that often only approximate complex phenomena. Neuromorphic atomic switch networks align their own innate structural complexity with that of the phenomenon they are modeling. They are also inherently fast—the state of the network can fluctuate at upward of tens of thousands of changes per second. "We are using a complex system to understand complex phenomena," Gimzewski said.

At a 2017 meeting of the American Chemical Society in San Francisco, Gimzewski, Stieg and their colleagues presented the results of an experiment in which they fed the device the first three years of a six-year data set of car traffic in Los Angeles, in the form of a series of pulses that indicated the number of cars passing by per hour. After hundreds of training runs, the output eventually predicted the statistical trend of the second half of the data set quite well, even though the device had never seen it.

Perhaps one day, Gimzewski jokes, he might be able to use the network to predict the stock market. "I'd like that," he said, adding that this was why he was trying to get his students to study atomic switch networks—"before they catch me making a fortune."

CLEVER MACHINES LEARN HOW TO BE CURIOUS

John Pavlus

You probably can't remember what it feels like to play Super Mario Bros. for the very first time, but try to picture it. An 8-bit game world blinks into being: baby blue sky, tessellated stone ground, and in between, a squat, red-suited man standing still—waiting. He's facing rightward; you nudge him farther in that direction. A few more steps reveal a row of bricks hovering overhead and what looks like an angry, ambulatory mushroom. Another twitch of the game controls makes the man spring up, his four-pixel fist pointed skyward. What now? Maybe try combining *nudge-rightward* and *spring-skyward?* Done. Then, a surprise: The little man bumps his head against one of the hovering bricks, which flexes upward and then snaps back down as if spring-loaded, propelling the man earthward onto the approaching angry mushroom and flattening it instantly. Mario bounces off the squished remains with a gentle hop. Above, copper-colored boxes with glowing "?" symbols seem to ask: What now?

This scene will sound familiar to anyone who grew up in the 1980s, but you can watch a much younger player on Pulkit Agrawal's YouTube channel. Agrawal, a computer science researcher at the University of California, Berkeley, is studying how innate curiosity can make learning an unfamiliar task—like playing Super Mario Bros. for the very first time—more efficient. The catch is that the novice player in Agrawal's video isn't human, or even alive. Like Mario, it's just software. But this software comes equipped with experimental machine-learning algorithms designed by Agrawal and his colleagues Deepak Pathak, Alexei A. Efros and Trevor Darrell at the Berkeley Artificial Intelligence Research Lab for a surprising purpose: to make a machine curious.[1]

"You can think of curiosity as a kind of reward which the agent generates internally on its own, so that it can go explore more about its world," Agrawal said. This internally generated reward signal is known in cognitive psychology as "intrinsic motivation." The feeling you may have vicariously experienced while reading the game-play description above—an urge to reveal more of whatever's waiting just out of sight, or just beyond your reach, just to see what happens—that's intrinsic motivation.

Humans also respond to extrinsic motivations, which originate in the environment. Examples of these include everything from the salary you receive at work to a demand delivered at gunpoint. Computer scientists apply a similar approach called reinforcement learning to train their algorithms: The software gets "points" when it performs a desired task, while penalties follow unwanted behavior.

But this carrot-and-stick approach to machine learning has its limits, and artificial intelligence researchers are starting to view intrinsic motivation as an important component of software agents that can learn efficiently and flexibly—that is, less like brittle machines and more like humans and animals. Approaches to using intrinsic motivation in AI have taken inspiration from psychology and neurobiology—not to mention decades-old AI research itself, now newly relevant. ("Nothing is really new in machine learning," said the artificial intelligence researcher Rein Houthooft.) ·

Such agents may be trained on video games now, but the impact of developing meaningfully "curious" AI would transcend any novelty appeal. "Pick your favorite application area and I'll give you an example," said Darrell, co-director of the Berkeley Artificial Intelligence lab. "At home, we want to automate cleaning up and organizing objects. In logistics, we want inventory to be moved around and manipulated. We want vehicles that can navigate complicated environments and rescue robots that can explore a building and find people who need rescuing. In all of these cases, we are trying to figure out this really hard problem: How do you make a machine that can figure its own task out?"

THE PROBLEM WITH POINTS

Reinforcement learning is a big part of what helped Google's AlphaGo software beat the world's best human player at Go, an ancient and intuitive game long considered invulnerable to machine learning. The details of successfully using reinforcement learning in a particular domain are complex, but the general idea is simple: Give a learning algorithm, or "agent," a reward function, a mathematically defined signal to seek out and maximize. Then set it loose in an environment, which could be any real or virtual world. As the agent operates in the environment, actions that increase the value of the reward function get reinforced. With enough repetition—and if there's anything that computers are better at than people, it's repetition—the agent learns patterns of action, or policies, that maximize its reward function. Ideally, these policies will result in the agent reaching some desirable end state (like "win at Go"), without a programmer or engineer having to hand-code every step the agent needs to take along the way.

In other words, a reward function is the guidance system that keeps a reinforcement-learning-powered agent locked on target. The more clearly that target is defined, the better the agent performs—that is why many of them are currently tested on old video games, which often provide simple extrinsic reward schemes based on points. (The blocky, two-dimensional graphics are useful, too: Researchers can run and repeat their experiments quickly because the games are relatively simple to emulate.)

Yet "in the real world, there are no points," said Agrawal. Computer scientists want to have their creations explore novel environments that don't come preloaded with quantifiable objectives.

In addition, if the environment doesn't supply extrinsic rewards quickly and regularly enough, the agent "has no clue whether it's doing something right or wrong," Houthooft said. Like a heat-seeking missile unable to lock onto a target, "it doesn't have any way of [guiding itself through] its environment, so it just goes haywire."

Moreover, even painstakingly defined extrinsic reward functions that can guide an agent to display impressively intelligent behavior—like AlphaGo's ability to best the world's top human Go player—won't easily transfer or generalize to any other context without extensive modification. And that work must be done by hand, which is precisely the kind of labor that machine learning is supposed to help us sidestep in the first place.

Instead of a battery of pseudo-intelligent agents that can reliably hit specified targets like those missiles, what we really want from AI is more like an internal piloting ability. "You make your own rewards, right?" Agrawal said. "There's no god constantly telling you 'plus one' for doing this or 'minus one' for doing that."

CURIOSITY AS CO-PILOT

Deepak Pathak never set out to model anything as airily psychological as curiosity in code. "The word 'curiosity' is nothing but saying, 'a model which leads an agent to efficiently explore its environment in the presence of noise,'" said Pathak, a researcher in Darrell's lab at Berkeley and the lead author of the recent work.

But in 2016, Pathak was interested in the sparse-rewards problem for reinforcement learning. Deep-learning software, powered by reinforcement learning techniques, had recently made significant gains in playing simple score-driven Atari games like *Space Invaders* and *Breakout*. But even slightly more complex games like *Super Mario Bros.*—which require navigating toward a goal distant in time and space without constant rewards, not to mention an ability to learn and successfully execute composite moves like running and jumping at the same time—were still beyond an AI's grasp.

Pathak and Agrawal, working with Darrell and Efros, equipped their learning agent with what they call an intrinsic curiosity module (ICM) designed to pull it forward through the game without going haywire (to borrow Houthooft's term). The agent, after all, has absolutely no prior understanding of how to play *Super Mario Bros.*—in fact, it's less like a novice player and more like a newborn infant.

Indeed, Agrawal and Pathak took inspiration from the work of Alison Gopnik and Laura Schulz, developmental psychologists at Berkeley and at the Massachusetts Institute of Technology, respectively, who showed that babies and toddlers are naturally drawn to play with objects that surprise them the most, rather than with objects that are useful to achieving some extrinsic goal. "One way to [explain] this kind of curiosity in children is that they build a model of what they know about the world, and then they conduct experiments to learn more about what they don't know," Agrawal said. These "experiments" can be anything that generates an outcome which the agent (in this case, an infant) finds unusual or unexpected. The child might start with random limb movements that cause new sensations (known as "motor babbling"), then progress up to more coordinated behaviors like chewing on a toy or knocking over a pile of blocks to see what happens.

In Pathak and Agrawal's machine-learning version of this surprise-driven curiosity, the AI first mathematically represents what the current video frame of *Super Mario Bros.* looks like. Then it predicts what the game will look like several frames hence. Such a feat is well within the powers of current deep-learning systems. But then Pathak and Agrawal's ICM does something more. It generates an intrinsic reward signal defined by how wrong this prediction model turns out to be. The higher the error rate—that is, the more surprised it is—the higher the value of its intrinsic reward function. In other words, if a surprise is equivalent to noticing when something doesn't turn out as expected—that is, to being wrong—then Pathak and Agrawal's system gets rewarded for being surprised.

This internally generated signal draws the agent toward unexplored states in the game: informally speaking, it gets curious about what it doesn't yet know. And as the agent learns—that is, as its prediction model becomes less and less wrong—its reward signal from the ICM decreases, freeing the agent up to maximize the reward signal by exploring other, more surprising situations. "It's a way to make exploration go faster," Pathak said.

This feedback loop also allows the AI to quickly bootstrap itself out of a nearly blank-slate state of ignorance. At first, the agent is curious about any basic movement available to its onscreen body: Pressing right nudges Mario to the right, and then he stops; pressing right several times in a row makes Mario move without immediately stopping; pressing up makes him spring into the air, and then come down again; pressing down has no effect. This simulated motor babbling quickly converges on useful actions that move the agent forward into the game, even though the agent doesn't know it.

For example, since pressing down always has the same effect— nothing—the agent quickly learns to perfectly predict the effect of that action, which cancels the curiosity-supplied reward signal associated with it. Pressing up, however, has all kinds of unpredictable effects: Sometimes Mario goes straight up, sometimes in an arc; sometimes he takes a short hop, other times a long jump; sometimes he doesn't come down again (if, say, he happens to land on top of an obstacle). All of these outcomes register as errors in the agent's prediction model, resulting in a reward signal from the ICM, which makes the agent keep experimenting with that action. Moving to the right (which almost always reveals more game world) has similar curiosity-engaging effects. The impulse to move up and to the right can clearly be seen in Agrawal's demo video: Within seconds, the AI-controlled Mario starts hopping rightward like a hyperactive toddler, causing ever-more-unpredictable effects (like bumping against a hovering brick, or accidentally squishing a mushroom), all of which drive further exploration.

"By using this curiosity, the agent learns how to do all the things it needs to explore the world, like jump and kill enemies," explained Agrawal. "It doesn't even get penalized for dying. But it learns to avoid dying, because not-dying maximizes its exploration. It's reinforcing itself, not getting reinforcement from the game."

AVOIDING THE NOVELTY TRAP

Artificial curiosity has been a subject of AI research since at least the early 1990s. One way of formalizing curiosity in software centers on novelty-seeking: The agent is programmed to explore unfamiliar states in its environment. This broad definition seems to capture an intuitive understanding of the experience of curiosity, but in practice it can cause the agent to become trapped in states that satisfy its built-in incentive but prevent any further exploration.

For example, imagine a television displaying nothing but static on its screen. Such a thing would quickly engage the curiosity of a purely novelty-seeking agent, because a square of randomly flickering visual noise is, by definition, totally unpredictable from one moment to the next. Since every pattern of static appears entirely novel to the agent, its intrinsic reward function will ensure that it can never cease paying attention to this single, useless feature of the environment—and it becomes trapped.

It turns out that this type of pointless novelty is ubiquitous in the kind of richly featured environments—virtual or physical—that AI must learn to cope with to become truly useful. For example, a self-driving delivery vehicle equipped with a novelty-seeking intrinsic reward function might never make it past the end of the block. "Say you're moving along a street and the wind is blowing and the leaves of a tree are moving," Agrawal said. "It's very, very hard to predict where every leaf is going to go. If you're predicting pixels, these kinds of interactions will cause you to have high prediction errors, and make you very curious. We want to avoid that."

Agrawal and Pathak had to come up with a way to keep their agent curious, but not too curious. Predicting pixels—that is, using deep learning and computer vision to model an agent's visual field in its entirety from moment to moment—makes it hard to filter out potential distractions. It's computationally expensive, too.

So instead, the Berkeley researchers engineered their Mario-playing agent to translate its visual input from raw pixels into an abstracted version of reality. This abstraction incorporates only features of the environment that have the potential to affect the agent (or that the agent can influence). In essence, if the agent can't interact with a thing, it won't even be perceived in the first place.

Using this stripped-down "feature space" (versus the unprocessed "pixel space") not only simplifies the agent's learning process, it also neatly sidesteps the novelty trap. "The agent can't get any benefit out of modeling, say, clouds moving overhead, to predict the effects of its actions," explained Darrell. "So it's just not going to pay attention to the clouds when it's being curious. The previous versions of curiosity—at least some of them—were really only considering pixel-level prediction. Which is great, except for when you suddenly pass a very unpredictable but very boring thing."

THE LIMITS OF ARTIFICIAL CURIOSITY

Darrell conceded that this model of curiosity isn't perfect. "The system learns what's relevant, but there's no guarantee it'll always get it right," he said. Indeed, the agent makes it only about halfway through the first level of *Super*

Mario Bros. before getting trapped in its own peculiar local optimum. "There's this big gap which the agent has to jump across, which requires executing 15 or 16 continuous actions in a very, very specific order," Agrawal said. "Because it is never able to jump this gap, it dies every time by going there. And when it learns to perfectly predict this outcome, it stops becoming curious about going any further in the game." (In the agent's defense, Agrawal notes that this flaw emerges because the AI can press its simulated directional controls only in discrete intervals, which makes certain moves impossible.)

Ultimately, the problem with artificial curiosity is that even researchers who have studied intrinsic motivation for years still can't precisely define what curiosity is. Paul Schrater, a neuroscientist who leads the Computational Perception and Action Lab at the University of Minnesota, said that the Berkeley model "is the most intelligent thing to do in the short term to get an agent to automatically learn a novel environment," but he thinks it has less to do with "the intuitive concept of curiosity" than with motor learning and control. "It's controlling things that are beneath cognition, and more in the details of what the body does," he said.

To Schrater, the Berkeley team's novel idea comes in attaching their intrinsic curiosity module to an agent that perceives *Super Mario Bros.* as a feature space rather than as sequential frames of pixels. He argues that this approach may roughly approximate the way our own brains "extract visual features that are relevant for a particular kind of task."

Curiosity may also require an agent to be at least somewhat embodied (virtually or physically) within an environment to have any real meaning, said Pierre-Yves Oudeyer, a research director at Inria in Bordeaux, France. Oudeyer has been creating computational models of curiosity for over a decade. He pointed out that the world is so large and rich that an agent can find surprises everywhere. But this isn't itself enough. "If you've got a disembodied agent using curiosity to explore a large feature space, its behavior is going to just end up looking like random exploration because it doesn't have any constraints on its actions," Oudeyer said. "The constraints of, for example, a body enable a simplification of the world." They focus the attention and help to guide exploration.

But not all embodied agents need intrinsic motivation, either—as the history of industrial robotics makes clear. For tasks that are simpler to specify—say, shuttling cargo from place to place using a robot that follows a yellow line painted on the floor—adding curiosity to the mix would be machine-learning overkill.

"You could just give that kind of agent a perfect reward function— everything it needs to know in advance," Darrell explained. "We could solve that problem 10 years ago. But if you're putting a robot in a situation that can't be modeled in advance, like disaster search-and-rescue, it has to go out and learn to explore on its own. That's more than just mapping—it has to learn the effects of its own actions in the environment. You definitely want an agent to be curious when it's learning how to do its job."

AI is often informally defined as "whatever computers can't do yet." If intrinsic motivation and artificial curiosity are methods for getting agents to figure out tasks that we don't already know how to automate, then "that's something I'm pretty sure we'd want any AI to have," said Houthooft. "The difficulty is in tuning it." Agrawal and Pathak's Mario-playing agent may not be able to get past World 1–1 on its own. But that's probably what tuning curiosity—artificial or otherwise—will look like: a series of baby steps.

VII HOW WILL WE LEARN MORE?

GRAVITATIONAL WAVES DISCOVERED AT LONG LAST

Natalie Wolchover

R ipples in space-time caused by the violent mergers of black holes have been detected, 100 years after these "gravitational waves" were predicted by Albert Einstein's theory of general relativity and half a century after physicists set out to look for them.

The landmark discovery was reported in February 2016 by the Advanced Laser Interferometer Gravitational-Wave Observatory (Advanced LIGO) team, confirming months of rumors that had surrounded the group's analysis of its first round of data. Astrophysicists say the detection of gravitational waves opens up a new window on the universe, revealing faraway events that can't be seen by optical telescopes, but whose faint tremors can be felt, even heard, across the cosmos.

"We have detected gravitational waves. We did it!" announced David Reitze, executive director of the 1,000-member team, at a National Science Foundation press conference in Washington, D.C.

Gravitational waves are perhaps the most elusive prediction of Einstein's theory, one that he and his contemporaries debated for decades. According to his theory, space and time form a stretchy fabric that bends under heavy objects, and to feel gravity is to fall along the fabric's curves. But can the "space-time" fabric ripple like the skin of a drum? Einstein flip-flopped, confused as to what his equations implied. But even steadfast believers assumed that, in any case, gravitational waves would be too weak to observe. They cascade outward from certain cataclysmic events, alternately stretching and squeezing space-time as they go. But by the time the waves reach Earth from these remote sources, they typically stretch and squeeze each mile of space by a minuscule fraction of the width of an atomic nucleus.

Perceiving the waves took patience and a delicate touch. Advanced LIGO bounced laser beams back and forth along the four-kilometer arms of two L-shaped detectors—one in Hanford, Washington, the other in Livingston,

Louisiana—looking for coincident expansions and contractions of their arms caused by gravitational waves as they passed. Using state-of-the-art stabilizers, vacuums and thousands of sensors, the scientists measured changes in the arms' lengths as tiny as one thousandth the width of a proton. This sensitivity would have been unimaginable a century ago, and struck many as implausible in 1968, when Rainer Weiss of the Massachusetts Institute of Technology conceived the experiment that became LIGO.

"The great wonder is they did finally pull it off; they managed to detect these little boogers!" said Daniel Kennefick, a theoretical physicist at the University of Arkansas and author of the 2007 book *Traveling at the Speed of Thought: Einstein and the Quest for Gravitational Waves.*

The detection ushers in a new era of gravitational-wave astronomy that is expected to deliver a better understanding of the formation, population and galactic role of black holes—super-dense balls of mass that curve space-time so steeply that even light cannot escape. When black holes spiral toward each other and merge, they emit a "chirp": space-time ripples that grow higher in pitch and amplitude before abruptly ending. The chirps that LIGO can detect happen to fall in the audible range, although they are far too quiet to be heard by the unaided ear. You can re-create the sound by running your finger along a piano's keys. "Start from the lowest note on the piano and go to middle C," Weiss said. "That's what we hear."

Physicists are already surprised by the number and strength of the signals detected so far, which imply that there are more black holes out there than expected. "We got lucky, but I was always expecting us to be somewhat lucky," said Kip Thorne, a theoretical physicist at the California Institute of Technology who founded LIGO with Weiss and the late Ronald Drever, who was also at Caltech. "This usually happens when a whole new window's been opened up on the universe."

Eavesdropping on gravitational waves could reshape our view of the cosmos in other ways, perhaps uncovering unimagined cosmic happenings.

"I liken this to the first time we pointed a telescope at the sky," said Janna Levin, a theoretical astrophysicist at Barnard College of Columbia University. "People realized there was something to see out there, but didn't foresee the huge, incredible range of possibilities that exist in the universe." Similarly, Levin said, gravitational-wave detections might possibly reveal that "the universe is full of dark stuff that we simply can't detect in a telescope."

The story of the first gravitational-wave detection began on a Monday morning in September 2015, and it started with a bang: a signal so loud and clear that Weiss thought, "This is crap. It's gotta be no good."

FEVER PITCH

That first gravitational wave swept across Advanced LIGO's detectors—first at Livingston, then at Hanford seven milliseconds later—during a mock run in the early hours of Sept. 14, 2015, two days before data collection was officially scheduled to begin.

The detectors were just firing up again after a five-year, $200-million upgrade, which equipped them with new noise-damping mirror suspensions and an active feedback system for canceling out extraneous vibrations in real time. The upgrades gave Advanced LIGO a major sensitivity boost over its predecessor, "initial LIGO," which from 2002 to 2010 had detected "a good clean zero," as Weiss put it.

When the big signal arrived that September, scientists in Europe, where it was morning, frantically emailed their American colleagues. As the rest of the team awoke, the news quickly spread. According to Weiss, practically everyone was skeptical—especially when they saw the signal. It was such a textbook chirp that many suspected the data had been hacked.

Mistaken claims in the search for gravitational waves have a long history, starting in the late 1960s when Joseph Weber of the University of Maryland thought he observed aluminum bars resonating in response to the waves. Most recently, in 2014, an experiment called BICEP2 reported the detection of primordial gravitational waves—space-time ripples from the Big Bang that would now be stretched and permanently frozen into the geometry of the universe. The BICEP2 team went public with great fanfare before their results were peer-reviewed, and then got burned when their signal turned out to have come from space dust.

When Lawrence Krauss, a cosmologist at Arizona State University, got wind of the Advanced LIGO detection, "the first thought is that it was a blind injection," he said. During initial LIGO, simulated signals had been secretly inserted into the data streams to test the response, unbeknownst to most of the team. When Krauss heard from an inside source that it wasn't a blind injection this time, he could hardly contain his excitement.

On Sept. 25, 2015, he tweeted to his 200,000 followers: "Rumor of a gravitational wave detection at LIGO detector. Amazing if true. Will post details if it survives." Then, on Jan. 11, 2016: "My earlier rumor about LIGO has been confirmed by independent sources. Stay tuned! Gravitational waves may have been discovered!"

The team's official stance was to keep quiet about their signal until they were dead sure. Thorne, bound by a vow of secrecy, didn't even tell his wife. "I celebrated in private," he said. The team's first step was to go back and analyze in excruciating detail how the signal had propagated through the detectors' thousands of different measurement channels, and to see whether anything strange had happened at the moment the signal was seen. They found nothing unusual. They also ruled out hackers, who would have had to know more than anyone about the experiment's thousands of data streams. "Even the team that does the blind injections have not perfected their injections well enough not to leave behind lots of fingerprints," Thorne said. "And there were no fingerprints."

Another, weaker chirp showed up in the weeks that followed.

The scientists analyzed these first two signals as even more swept in, and they submitted their paper to *Physical Review Letters*.[1] Their estimate of the statistical significance of the first, biggest signal is above "5-sigma," meaning the scientists are 99.9999 percent sure it's real.

LISTENING FOR GRAVITY

Einstein's equations of general relativity are so complex that it took 40 years for most physicists to agree that gravitational waves exist and are detectable—even in theory.

Einstein first thought that objects cannot shed energy in the form of gravitational radiation, then changed his mind. He showed in a seminal 1918 paper which ones could: Dumbbell-like systems that rotate about two axes at once, such as binary stars and supernovas popping like firecrackers, can make waves in space-time.

Still, Einstein and his colleagues continued to waffle. Some physicists argued that even if the waves exist, the world will oscillate with them and they cannot be felt. It wasn't until 1957 that Richard Feynman put that question to rest, with a thought experiment demonstrating that, if gravitational waves exist, they are theoretically detectable. But nobody knew how common those dumbbell-like sources might be in our cosmic neighborhood, or how strong or weak the resulting waves would be. "There was that ultimate question of: Will we ever really detect them?" Kennefick said.

In 1968, "Rai" Weiss was a young professor at MIT who had been roped into teaching a class on general relativity—a theory that he, as an experimentalist, knew little about—when news broke that Joseph Weber had detected gravitational waves. Weber had set up a trio of desk-size aluminum bars in two different U.S. states, and he reported that gravitational waves had set them all ringing.

Weiss' students asked him to explain gravitational waves and weigh in about the news. Looking into it, he was intimidated by the complex mathematics.

"I couldn't figure out what the hell [Weber] was doing—how the bar inter-acted with the gravitational wave." He sat for a long time, asking himself, "What's the most primitive thing I can think of that will detect gravitational waves?" An idea came to him that he calls the "conceptual basis of LIGO."

Imagine three objects sitting in space-time—say, mirrors at the corners of a triangle. "Send light from one to the other," Weiss said. "Look at the time it takes to go from one mass to another, and see if the time has changed." It turns out, he said, "you can do that quickly. I gave it to [my students] as a problem. Virtually the whole class was able to do that calculation."

In the next few years, as other researchers tried and failed to replicate the results of Weber's resonance-bar experiments (what he observed remains unclear, but it wasn't gravitational waves), Weiss began plotting a much more precise and ambitious experiment: a gravitational-wave interferome-ter.[2] Laser light would bounce between three mirrors in an L-shaped arrange-ment, forming two beams. The spacing of the peaks and troughs of the light waves would precisely measure the lengths of the two arms, creating what could be thought of as x and y axes for space-time. When the grid was still, the two light waves would bounce back to the corner and cancel each other out, producing a null signal in a detector. But if a gravitational wave swept across Earth, it would stretch the length of one arm and compress the length of the other (and vice versa in an alternating pattern). The off-alignment of the two light beams would create a signal in the detector, revealing a fleeting tremor in space and time.

Fellow physicists were skeptical at first, but the experiment soon found a champion in Thorne, whose theory group at Caltech studied black holes and other potential gravitational-wave sources and the signals they would produce. Thorne had been inspired by Weber's experiment and similar efforts by Russian physicists; after speaking with Weiss at a conference in 1975, "I began to believe that gravitational-wave detection would succeed," Thorne said, "and I wanted Caltech to be involved." He had Caltech hire the Scottish experimentalist Ronald Drever, who had also been clamoring to build a gravitational-wave interferometer. Thorne, Drever and Weiss eventually began working as a team, each taking on a share of the countless problems that had to be solved to develop a feasible experiment. The trio founded LIGO in 1984, and, after building prototypes and collaborating with a growing team, banked more than $100 million in NSF funding in the early 1990s. Blueprints were drawn up for a pair of giant L-shaped detectors. A decade later, the detectors went online.

In Hanford and Livingston, vacuums run down the center of each detec-tor's four-kilometer arms, keeping the laser, the beam path and the mirrors as isolated as possible from the planet's constant trembling. Not taking any chances, LIGO scientists monitor their detectors with thousands of instruments during each data run, measuring everything they can: seismic activity, atmospheric pressure, lightning, the arrival of cosmic rays, vibrations of the equipment, sounds near the laser beam and so on. They then cleanse their data of these various sources of background noise. Perhaps most importantly, having two detectors allows them to cross-check their data, looking for coincident signals.

Inside the vacuum, even with isolated and stabilized lasers and mirrors, "strange signals happen all the time," said Marco Cavaglià, then assistant spokesperson for the LIGO collaboration, in 2016. The scientists must trace these "koi fish," "ghosts," "fringy sea monsters" and other rogue vibrational patterns back to their sources so the culprits can be removed. One tough case occurred during the testing phase, said Jessica McIver, a postdoctoral researcher and one of the team's foremost glitch detectives. It was a string of periodic, single-frequency artifacts that appeared every so often in the data. When she and her colleagues converted the mirror vibrations into an audio file, "you could clearly hear the ring-ring-ring of a telephone," McIver said. "It turned out to be telemarketers calling the phone inside the laser enclosure."

The sensitivity of Advanced LIGO's detectors will continue to improve in the coming years, and a third interferometer called Advanced Virgo has since come online in Italy. One question the data might help answer is how black holes form. Are they products of implosions of the earliest, massive stars, or do they originate from collisions inside tight clusters of stars? "Those are just two ideas; I bet there will be several more before the dust settles," Weiss said. As LIGO tallies new statistics in future runs, scientists will be listening for whispers of these black-hole origin stories.

Judging by its shape and size, that first, loudest chirp originated about 1.3 billion light-years away from the location where two black holes, each of roughly 30 solar masses, finally merged after slow-dancing under mutual gravitational attraction for eons. The black holes spiraled toward each other faster and faster as the end drew near, like water in a drain, shedding three suns' worth of energy to gravitational waves in roughly the blink of an eye. The merger is the most energetic event ever detected.

"It's as though we had never seen the ocean in a storm," Thorne said. He has been waiting for a storm in space-time ever since the 1960s. The feeling he experienced when the waves finally rolled in wasn't excitement, he said, but something else: profound satisfaction.

COLLIDING BLACK HOLES TELL NEW STORY OF STARS

Natalie Wolchover

A t an August 2016 talk in Santa Barbara, California, addressing some of the world's leading astrophysicists, Selma de Mink cut to the chase. "How did they form?" she began.

"They," as everybody knew, were the two massive black holes that, more than 1 billion years ago and in a remote corner of the cosmos, spiraled together and merged, making waves in the fabric of space and time. These "gravitational waves" rippled outward and, on Sept. 14, 2015, swept past Earth, strumming the ultrasensitive detectors of the Laser Interferometer Gravitational-Wave Observatory (LIGO). LIGO's discovery, announced in February 2016, triumphantly vindicated Albert Einstein's 1916 prediction that gravitational waves exist. By tuning in to these tiny tremors in space-time and revealing for the first time the invisible activity of black holes—objects so dense that not even light can escape their gravitational pull—LIGO promised to open a new window on the universe, akin, some said, to when Galileo first pointed a telescope at the sky.

The new gravitational-wave data has shaken up the field of astrophysics. In response, three dozen experts spent two weeks in August 2016 sorting through the implications at the Kavli Institute for Theoretical Physics (KITP) in Santa Barbara.

Jump-starting the discussions, de Mink, an astrophysicist at the University of Amsterdam, explained that of the two—and possibly more—black-hole mergers that LIGO has detected so far, the first and mightiest event, labeled GW150914, presented the biggest puzzle. LIGO was expected to spot pairs of black holes weighing in the neighborhood of 10 times the mass of the sun, but these packed roughly 30 solar masses apiece. "They are there— massive black holes, much more massive than we thought they were," de Mink said to the room. "So, how did they form?"

The mystery, she explained, is twofold: How did the black holes get so massive, considering that stars, some of which collapse to form black holes, typically blow off most of their mass before they die, and how did they get so close to each other—close enough to merge within the lifetime of the universe? "These are two things that are sort of mutually exclusive," de Mink said. A pair of stars that are born huge and close together will normally mingle and then merge before ever collapsing into black holes, failing to kick up detectable gravitational waves.

Nailing down the story behind GW150914 "is challenging all our understanding," said the astrophysicist Matteo Cantiello. Experts must retrace the uncertain steps from the moment of the merger back through the death, life and birth of a pair of stars—a sequence that involves much unresolved astrophysics. "This will really reinvigorate certain old questions in our understanding of stars," said Eliot Quataert, a professor of astronomy at the University of California, Berkeley, and one of the organizers of the KITP program. Understanding LIGO's data will demand a reckoning of when and why stars go supernova; which ones turn into which kinds of stellar remnants; how stars' composition, mass and rotation affect their evolution; how their magnetic fields operate; and more.

The work has just begun, but already LIGO's first detections have pushed two theories of binary black-hole formation to the front of the pack. Over the two weeks in Santa Barbara, a rivalry heated up between the new "chemically homogeneous" model for the formation of black-hole binaries, proposed by de Mink and colleagues in 2016, and the classic "common envelope" model espoused by many other experts. Both theories (and a cluster of competitors) might be true somewhere in the cosmos, but probably only one of them accounts for the vast majority of black-hole mergers. "In science," said Daniel Holz of the University of Chicago, a common-envelope proponent, "there's usually only one dominant process—for anything."

STAR STORIES

The story of GW150914 almost certainly starts with massive stars—those that are at least eight times as heavy as the sun and which, though rare, play a starring role in galaxies. Massive stars are the ones that explode as supernovas, spewing matter into space to be recycled as new stars; only their cores then collapse into black holes and neutron stars, which drive exotic and influential phenomena such as gamma-ray bursts, pulsars and X-ray binaries. De Mink and collaborators showed in 2012 that most known massive stars live in binary systems.[1] Binary massive stars, in her telling, "dance" and "kiss" and suck each other's hydrogen fuel "like vampires," depending on the circumstances. But which circumstances lead them to shrink down to points that recede behind veils of darkness, and then collide?

The conventional common-envelope story, developed over decades starting with the 1970s work of the Soviet scientists Aleksandr Tutukov and Lev Yungelson, tells of a pair of massive stars that are born in a wide orbit. As the first star runs out of fuel in its core, its outer layers of hydrogen puff up, forming a "red supergiant." Much of this hydrogen gas gets sucked away by the second star, vampire-style, and the core of the first star eventually collapses into a black hole. The interaction draws the pair closer, so that when the second star puffs up into a supergiant, it engulfs the two of them in a common envelope. The companions sink ever closer as they wade through the hydrogen gas. Eventually, the envelope is lost to space, and the core of the second star, like the first, collapses into a black hole. The two black holes are close enough to someday merge.

Because the stars shed so much mass, this model is expected to yield pairs of black holes on the lighter side, weighing in the ballpark of 10 solar masses. LIGO's second signal, from the merger of eight- and 14-solar-mass black holes, is a home run for the model. But some experts say that the first event, GW150914, is a stretch.

In a June 2016 paper in *Nature*, Holz and collaborators Krzysztof Belczynski, Tomasz Bulik and Richard O'Shaughnessy argued that common envelopes can theoretically produce mergers of 30-solar-mass black holes if the progenitor stars weigh something like 90 solar masses and contain almost no metal (which accelerates mass loss).[2] Such heavy binary systems are likely to be relatively rare in the universe, raising doubts in some minds about whether LIGO would have observed such an outlier so soon. In Santa Barbara, scientists agreed that if LIGO detects many very heavy mergers relative to lighter ones, this will weaken the case for the common-envelope scenario.

This weakness of the conventional theory has created an opening for new ideas. One such idea began brewing in 2014, when de Mink and Ilya Mandel, an astrophysicist at the University of Birmingham and a member of the LIGO collaboration, realized that a type of binary-star system that de Mink has studied for years might be just the ticket to forming massive binary black holes.

The chemically homogeneous model begins with a pair of massive stars that are rotating around each other extremely rapidly and so close together that they become "tidally locked," like tango dancers. In tango, "you are extremely close, so your bodies face each other all the time," said de Mink, a dancer herself. "And that means you are spinning around each other, but it also forces you to spin around your own axis as well." This spinning stirs the stars, making them hot and homogeneous throughout. And this process might allow the stars to undergo fusion throughout their whole interiors, rather than just their cores, until both stars use up all their fuel. Because the stars never expand, they do not intermingle or shed mass. Instead, each collapses wholesale under its own weight into a massive black hole. The black holes dance for a few billion years, gradually spiraling closer and closer until, in a space-time-buckling split second, they coalesce.

De Mink and Mandel made their case for the chemically homogeneous model in a paper posted online in January 2016.[3] Another paper proposing the same idea, by researchers at the University of Bonn led by the graduate student Pablo Marchant, appeared days later.[4] When LIGO announced the detection of GW150914 the following month, the chemically homogeneous theory shot to prominence. "What I'm discussing was a pretty crazy story up to the moment that it made, very nicely, black holes of the right mass," de Mink said.

However, aside from some provisional evidence, the existence of stirred stars is speculative. And some experts question the model's efficacy. Simulations suggest that the chemically homogeneous model struggles to explain smaller black-hole binaries like those in LIGO's second signal. Worse, doubt has arisen as to how well the theory really accounts for GW150914, which is supposed to be its main success story. "It's a very elegant model," Holz said. "It's very compelling. The problem is that it doesn't seem to fully work."

ALL SPUN UP

Along with the masses of the colliding black holes, LIGO's gravitational-wave signals also reveal whether the black holes were spinning. At first, researchers paid less attention to the spin measurement, in part because gravitational waves only register spin if black holes are spinning around the same axis that they orbit each other around, saying nothing about spin in other directions. However, in a May 2016 paper, researchers at the Institute for Advanced Study in Princeton, New Jersey, and the Hebrew University of Jerusalem argued that the kind of spin that LIGO measures is exactly the kind black holes would be expected to have if they formed via the chemically homogeneous channel.[5] (Tango dancers spin and orbit each other in the same direction.) And yet, the 30-solar-mass black holes in GW150914 were measured to have very low spin, if any, seemingly striking a blow against the tango scenario.

"Is spin a problem for the chemically homogeneous channel?" Sterl Phinney, a professor of astrophysics at the California Institute of Technology, prompted the Santa Barbara group one afternoon. After some debate, the scientists agreed that the answer was yes.

However, mere days later, de Mink, Marchant and Cantiello found a possible way out for the theory. Cantiello, who has recently made strides in studying stellar magnetic fields, realized that the tangoing stars in the chemically homogeneous channel are essentially spinning balls of charge that would have powerful magnetic fields, and these magnetic fields are likely to cause the star's outer layers to stream into strong poles. In the same way that a spinning figure skater slows down when she extends her arms, these poles would act like brakes, gradually reducing the stars' spin. The trio has since been working to see if their simulations bear out this picture. Quataert called the idea "plausible but perhaps a little weaselly."

On the last day of the program, which set the stage for an eventful autumn as LIGO came back online with higher sensitivity and more gravitational-wave signals rolling in, the scientists signed "Phinney's Declaration," a list of concrete statements about what their various theories predict. "Though all models for black hole binaries may be created equal (except those inferior ones proposed by our competitors)," begins the declaration, drafted by Phinney, "we hope that observational data will soon make them decidedly unequal."

As the data pile up, an underdog theory of black-hole binary formation could conceivably gain traction—for instance, the notion that binaries form through dynamical interactions inside dense star-forming regions called "globular clusters." LIGO's first run suggested that black-hole mergers are more common than the globular-cluster model predicts. But perhaps the experiment just got lucky last time and the estimated merger rate will drop.

Adding to the mix, a group of cosmologists theorized that GW150914 might have come from the merger of primordial black holes, which were never stars to begin with but rather formed shortly after the Big Bang from the collapse of energetic patches of space-time. Intriguingly, the researchers argued in a paper in *Physical Review Letters* that such 30-solar-mass primordial black holes could comprise some or all of the missing "dark matter" that pervades the cosmos.[6] There's a way of testing the idea against astrophysical signals called fast radio bursts.

It's perhaps too soon to dwell on such an enticing possibility; astrophysicists point out that it would require suspiciously good luck for black holes from the Big Bang to happen to merge at just the right time for us to detect them, 13.8 billion years later. This is another example of the new logic that researchers must confront at the dawn of gravitational-wave astronomy. "We're at a really fun stage," de Mink said. "This is the first time we're thinking in these pictures."

NEUTRON-STAR COLLISION SHAKES SPACE-TIME AND LIGHTS UP THE SKY

Katia Moskvitch

O n Aug. 17, 2017, the Advanced Laser Interferometer Gravitational-Wave Observatory (LIGO) detected something new. Some 130 million light-years away, two super-dense neutron stars, each as small as a city but heavier than the sun, had crashed into each other, producing a colossal convulsion called a kilonova and sending a telltale ripple through space-time to Earth.

When LIGO picked up the signal, the astronomer Edo Berger was in his office at Harvard University suffering through a committee meeting. Berger leads an effort to search for the afterglow of collisions detected by LIGO. But when his office phone rang, he ignored it. Shortly afterward, his cell phone rang. He glanced at the display to discover a flurry of missed text messages:

Edo, check your email!

Pick up your phone!

"I kicked everybody out that very moment and jumped into action," Berger said. "I had not expected this."

LIGO's pair of ultrasensitive detectors in Louisiana and Washington state made history in 2015 by recording the gravitational waves coming from the collision of two black holes—a discovery that earned the experiment's architects the Nobel Prize in Physics in 2017. More signals from black hole collisions followed the initial discovery.

Yet black holes don't give off light, so making any observations of these faraway cataclysms beyond the gravitational waves themselves was unlikely. Colliding neutron stars, on the other hand, produce fireworks. Astronomers had never seen such a show before, but now LIGO was telling them where to look, which sent teams of researchers like Berger's scurrying to capture the immediate aftermath of the collision across the full range of electromagnetic signals. In total, more than 70 telescopes swiveled toward the same location in the sky.

They struck the motherlode. In the days after the initial detection, astronomers made successful observations of the colliding neutron stars with optical, radio, X-ray, gamma-ray, infrared and ultraviolet telescopes. The enormous collaborative effort, detailed in dozens of papers appearing simultaneously in *Physical Review Letters, Nature, Science, Astrophysical Journal Letters* and other journals, has not only allowed astrophysicists to piece together a coherent account of the event, but also to answer longstanding questions in astrophysics.

"In one fell swoop, gravitational wave measurements" have opened "a window onto nuclear astrophysics, neutron star demographics and physics and precise astronomical distances," said Scott Hughes, an astrophysicist at the Massachusetts Institute of Technology's Kavli Institute for Astrophysics and Space Research. "I can't describe in family-friendly words how exciting that is."

The discovery, Berger said, "will go down in the history of astronomy."

X MARKS THE SPOT

When Berger got the calls, emails, and the automated official LIGO alert with the probable coordinates of what appeared to be a neutron-star merger, he knew that he and his team had to act quickly to see its aftermath using optical telescopes.

The timing was fortuitous. Virgo, a new gravitational-wave observatory similar to LIGO's two detectors, had just come online in Europe. The three gravitational-wave detectors together were able to triangulate the signal. Had the neutron-star merger occurred a month or two earlier, before Virgo started taking data, the "error box," or area in the sky that the signal could have come from, would have been so large that follow-up observers would have had little chance of finding anything.

The LIGO and Virgo scientists had another stroke of luck. Gravitational waves produced by merging neutron stars are fainter than those from black holes and harder to detect. According to Thomas Dent, an astrophysicist at the Albert Einstein Institute in Hannover, Germany, and a member of LIGO, the experiment can only sense neutron-star mergers that occur within 300 million light-years. This event was far closer—at a comfortable distance for both LIGO and the full range of electromagnetic telescopes to observe it.

But at the time, Berger and his colleagues didn't know any of that. They had an agonizing wait until sunset in Chile, when they could use an instrument called the Dark Energy Camera mounted on the Victor M. Blanco telescope there. The camera is great when you don't know precisely where you're looking, astronomers said, because it can quickly scan a very large area of the sky. Berger also secured use of the Very Large Array (VLA) in central New Mexico, the Atacama Large Millimeter Array (ALMA) in Chile and the space-based Chandra X-ray Observatory. (Other teams that received the LIGO alert asked to use VLA and ALMA as well.)

A few hours later, data from the Dark Energy Camera started coming in. It took Berger's team 45 minutes to spot a new bright light source. The light appeared to come from a galaxy called NGC 4993 in the constellation Hydra that had been pointed out in the LIGO alert, and at approximately the distance where LIGO had suggested they look.

"That got us really excited, and I still have the email from a colleague saying 'Holy [smokes], look at that bright source near this galaxy!'" Berger said. "All of us were kind of shocked," since "we didn't think we would succeed right away." The team had expected a long slog, maybe having to wade through multiple searches after LIGO detections for a couple of years until eventually spotting something. "But this just stood out," he said, "like when an X marks the spot."

Meanwhile, at least five other teams discovered the new bright light source independently, and hundreds of researchers made various follow-up observations. David Coulter, an astronomer at University of California, Santa Cruz, and colleagues used the Swope telescope in Chile to pinpoint the event's exact location, while Las Cumbres Observatory astronomers did so with the help of a robotic network of 20 telescopes around the globe.

For Berger and the rest of the Dark Energy Camera follow-up team, it was time to call in the Hubble Space Telescope. Securing time on the veteran instrument usually takes weeks, if not months. But for extraordinary circumstances, there's a way to jump ahead in line, by using "director's discretionary time." Matt Nicholl, an astronomer at the Harvard-Smithsonian Center for Astrophysics, submitted a proposal on behalf of the team to take ultraviolet measurements with Hubble—possibly the shortest proposal ever written. "It was two paragraphs long—that's all we could do in the middle of the night," Berger said. "It just said that we've found the first counterpart of a binary neutron star merger, and we need to get UV spectra. And it got approved."

As the data trickled in from the various instruments, the collected data set was becoming more and more astounding. In total, the original LIGO/Virgo discovery and the various follow-up observations by scientists have yielded dozens of papers, each describing astrophysical processes that occurred during and after the merger.

MYSTERY BURSTS

Neutron stars are compact neutron-packed cores left over when massive stars die in supernova explosions. A teaspoon of neutron star would weigh as much as one billion tons. Their internal structure is not completely understood. Neither is their occasional aggregation into close-knit binary pairs of stars that orbit each other. The astronomers Joe Taylor and Russell Hulse found the first such pair in 1974, a discovery that earned them the 1993 Nobel Prize in Physics. They concluded that those two neutron stars were destined to crash into each other in about 300 million years. The two stars newly discovered by LIGO took far longer to do so.

The analysis by Berger and his team suggests that the newly discovered pair was born 11 billion years ago, when two massive stars went supernova a few million years apart. Between these two explosions, something brought the stars closer together, and they went on circling each other for most of the history of the universe. The findings are "in excellent agreement with the models of binary-neutron-star formation," Berger said.

The merger also solved another mystery that has vexed astrophysicists for the past five decades.

On July 2, 1967, two United States satellites, Vela 3 and 4, spotted a flash of gamma radiation. Researchers first suspected a secret nuclear test conducted by the Soviet Union. They soon realized this flash was something else: the first example of what is now known as a gamma ray burst (GRB), an event lasting anywhere from milliseconds to hours that "emits some of the most intense and violent radiation of any astrophysical object," Dent said. The origin of GRBs has been an enigma, although some people have suggested that so-called "short" gamma-ray bursts (lasting less than two seconds) could be the result of neutron-star mergers. There was no way to directly check until now.

In yet another nod of good fortune, it so happened that on Aug. 17, 2017, the Fermi Gamma-Ray Space Telescope and the International Gamma-Ray Astrophysics Laboratory (Integral) were pointing in the direction of the constellation Hydra. Just as LIGO and Virgo detected gravitational waves, the gamma-ray space telescopes picked up a weak GRB, and, like LIGO and Virgo, issued an alert.

A neutron star merger should trigger a very strong gamma-ray burst, with most of the energy released in a fairly narrow beam called a jet. The researchers believe that the GRB signal hitting Earth was weak only because the jet was pointing at an angle away from us. Proof arrived about two weeks later, when observatories detected the X-ray and radio emissions that accompany a GRB. "This provides smoking-gun proof that normal short gamma-ray bursts are produced by neutron-star mergers," Berger said. "It's really the first direct compelling connection between these two phenomena."

Hughes said that the observations were the first in which "we have definitively associated any short gamma-ray burst with a progenitor." The findings indicate that at least some GRBs come from colliding neutron stars, though it's too soon to say whether they all do.

STRIKING GOLD

Optical and infrared data captured after the neutron-star merger also help clarify the formation of the heaviest elements in the universe, like uranium, platinum and gold, in what's called r-process nucleosynthesis. Scientists long believed that these rare, heavy elements, like most other elements, are made during high-energy events such as supernovas. A competing theory that has gained prominence in recent years argues that neutron-star mergers could forge the majority of these elements. According to that thinking, the crash of neutron stars ejects matter in what's called a kilonova. "Once released from the neutron stars' gravitational field," the matter "would transmute into a cloud full of the heavy elements we see on rocky planets like Earth," Dent explained.

Optical telescopes picked up the radioactive glow of these heavy elements—strong evidence, scientists say, that neutron-star collisions produce much of the universe's supply of heavy elements like gold.

"With this merger," Berger said, "we can see all the expected signatures of the formation of these elements, so we are solving this big open question in astrophysics of how these elements form. We had hints of this before, but here we have a really nearby object with exquisite data, and there is no ambiguity." According to Daniel Holz of the University of Chicago, "back-of-the-envelope calculations indicate that this single collision produced an amount of gold greater than the weight of the Earth."

The scientists also inferred a sequence of events that may have followed the neutron-star collision, providing insight into the stars' internal structure. Experts knew that the collision outcome "depends very much on how large the stars are and how 'soft' or 'springy'—in other words, how much they resist being deformed by super-strong gravitational forces," Dent said. If the stars are extra soft, they may immediately be swallowed up inside a newly formed black hole, but this would not leave any matter outside to produce a gamma-ray burst. "At the other end of the scale," he said, "the two neutron stars would merge and form an unstable, rapidly spinning super-massive neutron star, which could produce a gamma-ray burst after a holdup of tens or hundreds of seconds."

The most plausible case may lie somewhere in the middle: The two neutron stars may have merged into a doughnut-shaped unstable neutron star that launched a jet of super-energetic hot matter before finally collapsing as a black hole, Dent said.

Future observations of neutron-star mergers will settle these questions. And as the signals roll in, experts say the mergers will also serve as a precision tool for cosmologists. Comparing the gravitational-wave signal with the redshift, or stretching, of the electromagnetic signals offers a new way of measuring the so-called Hubble constant, which gives the age and expansion rate of the universe. Already, with this one merger, researchers were able to make an initial measurement of the Hubble constant "in a remarkably fundamental way, without requiring the multitude of assumptions" that go into estimating the constant by other methods, said Matthew Bailes, a member of the LIGO collaboration and a professor at the Swinburne University of Technology in Australia. Holz described the neutron star merger as a "standard siren" (in a nod to the term "standard candles" used for supernovas) and said that initial calculations suggest the universe is expanding at a rate of 70 kilometers per second per megaparsec, which puts LIGO's Hubble constant "smack in the middle of [previous] estimates."

To improve the measurement, scientists will have to spot many more neutron-star mergers. Given that LIGO and Virgo are still being fine-tuned to increase their sensitivity, Berger is optimistic. "It is clear that the rate of occurrence is somewhat higher than expected," he said. "By 2020 I expect at least one to two of these every month. It will be tremendously exciting."

WHERE DO WE GO FROM HERE?

WHAT NO NEW PARTICLES MEANS FOR PHYSICS

Natalie Wolchover

P hysicists at the Large Hadron Collider in Europe have explored the properties of nature at higher energies than ever before, and they have found something profound: nothing new.

It's perhaps the one thing that no one predicted 30 years ago when the project was first conceived.

The infamous "diphoton bump" that arose in data plots in December 2015 eventually disappeared, indicating that it was a fleeting statistical fluctuation rather than a revolutionary new fundamental particle. And in fact, the machine's collisions have so far conjured up no particles at all beyond those cataloged in the long-reigning but incomplete Standard Model of particle physics. In the collision debris, physicists have found no particles that could comprise dark matter, no siblings or cousins of the Higgs boson, no sign of extra dimensions, no leptoquarks—and above all, none of the desperately sought supersymmetry particles that would round out equations and satisfy the "naturalness" principle about how the laws of nature ought to work.

"It's striking that we've thought about these things for 30 years and we have not made one correct prediction that they have seen," said Nima Arkani-Hamed of the Institute for Advanced Study.

It was the summer of 2016, and the news emerged at the International Conference on High Energy Physics in Chicago in presentations by the ATLAS and CMS experiments, whose cathedral-like detectors sit at 6 and 12 o'clock on the LHC's 17-mile ring. Both teams, each with over 3,000 members, had been working feverishly for three months analyzing a glut of data from a machine that was finally running at full throttle, colliding protons with 13 trillion electron volts (TeV) of energy and providing enough raw material to beget gargantuan elementary particles, should any exist.

So far, none have materialized. Especially heartbreaking for many was the loss of the diphoton bump, an excess of pairs of photons that cropped up in a 2015 teaser batch of 13-TeV data, and whose origin has been the speculation of some 500 papers by theorists. Rumors about the bump's disappearance in 2016 data began leaking that June, triggering a community-wide "diphoton hangover."

"It would have single-handedly pointed to a very exciting future for particle experiments," said Raman Sundrum, a theoretical physicist at the University of Maryland. "Its absence puts us back to where we were."

The lack of new physics deepens a crisis that started in 2012 during the LHC's first run, when it became clear that its 8-TeV collisions would not generate any new physics beyond the Standard Model. (The Higgs boson, discovered that year, was the Standard Model's final puzzle piece, rather than an extension of it.) A white-knight particle could still show up later, or, as statistics accrue over a longer time scale, subtle surprises in the behavior of the known particles could indirectly hint at new physics. But theorists are increasingly bracing themselves for their "nightmare scenario," in which the LHC offers no path at all toward a more complete theory of nature.

Some theorists argue that the time has already come for the whole field to start reckoning with the message of the null results. The absence of new particles almost certainly means that the laws of physics are not natural in the way physicists long assumed they are. "Naturalness is so well-motivated," Sundrum said, "that its actual absence is a major discovery."

MISSING PIECES

The main reason physicists felt sure that the Standard Model could not be the whole story is that its linchpin, the Higgs boson, has a highly unnatural-seeming mass. In the equations of the Standard Model, the Higgs is coupled to many other particles. This coupling endows those particles with mass, allowing them in turn to drive the value of the Higgs mass to and fro, like competitors in a tug-of-war. Some of the competitors are extremely strong—hypothetical particles associated with gravity might contribute (or deduct) as much as 10 million billion TeV to the Higgs mass—yet somehow its mass ends up as 0.125 TeV, as if the competitors in the tug-of-war finish in a near-perfect tie. This seems absurd—unless there is some reasonable explanation for why the competing teams are so evenly matched.

Supersymmetry, as theorists realized in the early 1980s, does the trick. It says that for every "fermion" that exists in nature—a particle of matter, such as an electron or quark, that adds to the Higgs mass—there is a supersymmetric "boson," or force-carrying particle, that subtracts from the Higgs mass. This way, every participant in the tug-of-war game has a rival of equal strength, and the Higgs is naturally stabilized. Theorists devised alternative proposals for how naturalness might be achieved, but supersymmetry had additional arguments in its favor: It caused the strengths of the three quantum forces to exactly converge at high energies, suggesting they were unified at the beginning of the universe. And it supplied an inert, stable particle of just the right mass to be dark matter.

"We had figured it all out," said Maria Spiropulu, a particle physicist at the California Institute of Technology and a member of CMS. "If you ask people of my generation, we were almost taught that supersymmetry is there even if we haven't discovered it. We believed it."

Hence the surprise when the supersymmetric partners of the known particles didn't show up—first at the Large Electron-Positron Collider in the 1990s, then at the Tevatron in the 1990s and early 2000s, and now at the LHC. As the colliders have searched ever-higher energies, the gap has widened between the known particles and their hypothetical superpartners, which must be much heavier in order to have avoided detection. Ultimately, supersymmetry becomes so "broken" that the effects of the particles and their superpartners on the Higgs mass no longer cancel out, and supersymmetry fails as a solution to the naturalness problem. Some experts argued in 2016 that we'd passed that point already. Others, allowing for more freedom in how certain factors are arranged, said it was happening right then, with ATLAS and CMS excluding the stop quark—the hypothetical superpartner of the 0.173-TeV top quark—up to a mass of 1 TeV. That's already a nearly sixfold imbalance between the top and the stop in the Higgs tug-of-war. Even if a stop heavier than 1 TeV existed, it would be pulling too hard on the Higgs to solve the problem it was invented to address.·

"I think 1 TeV is a psychological limit," said Albert de Roeck, a senior research scientist at CERN, the laboratory that houses the LHC, and a professor at the University of Antwerp in Belgium.

Some will say that enough is enough, but for others there are still loopholes to cling to. Among the myriad supersymmetric extensions of the Standard Model, there are more complicated versions in which stop quarks heavier than 1 TeV conspire with additional supersymmetric particles to counterbalance the top quark, tuning the Higgs mass. The theory has so many variants, or individual "models," that killing it outright is almost impossible. Joe Incandela, a physicist at the University of California, Santa Barbara, who announced the discovery of the Higgs boson on behalf of the CMS collaboration in 2012, and who now leads one of the stop-quark searches, said, "If you see something, you can make a model-independent statement that you see something. Seeing nothing is a little more complicated."

The Standard Model

FERMIONS (matter) | BOSONS (force carriers)
  Quarks   Leptons | Gauge bosons   Higgs boson

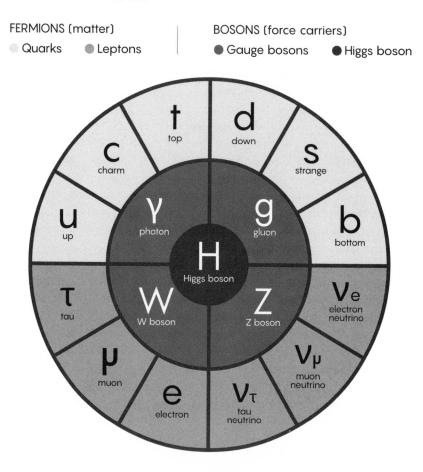

FIGURE 8.1

Particles can hide in nooks and crannies. If, for example, the stop quark and the lightest neutralino (supersymmetry's candidate for dark matter) happen to have nearly the same mass, they might have stayed hidden so far. The reason for this is that, when a stop quark is created in a collision and decays, producing a neutralino, very little energy will be freed up to take the form of motion. "When the stop decays, there's a dark-matter particle just kind of sitting there," explained Kyle Cranmer of New York University, a member of ATLAS. "You don't see it. So in those regions it's very difficult to look for." In that case, a stop quark with a mass as low as 0.6 TeV could still be hiding in the data.

Experimentalists will strive to close these loopholes in the coming years, or to dig out the hidden particles. Meanwhile, theorists who are ready to move on face the fact that they have no signposts from nature about which way to go. "It's a very muddled and uncertain situation," Arkani-Hamed said.

NEW HOPE

Many particle theorists now acknowledge a long-looming possibility: that the mass of the Higgs boson is simply unnatural—its small value resulting from an accidental, fine-tuned cancellation in a cosmic game of tug-of-war—and that we observe such a peculiar property because our lives depend on it. In this scenario, there are many, many universes, each shaped by different chance combinations of effects. Out of all these universes, only the ones with accidentally lightweight Higgs bosons will allow atoms to form and thus give rise to living beings. But this "anthropic" argument is widely disliked for being seemingly untestable.

In the past few years, some theoretical physicists have started to devise totally new natural explanations for the Higgs mass that avoid the fatalism of anthropic reasoning and do not rely on new particles showing up at the LHC. Theorists have put forward ideas such as the relaxation hypothesis—which supposes that the Higgs mass, rather than being shaped by symmetry, was sculpted dynamically by the birth of the cosmos—and possible ways to test these ideas.[1] Nathaniel Craig of the University of California, Santa Barbara, who works on an idea called "neutral naturalness," said in a phone call from the 2016 CERN workshop, "Now that everyone is past their diphoton hangover, we're going back to these questions that are really aimed at coping with the lack of apparent new physics at the LHC."[2]

Arkani-Hamed, who, along with several colleagues, recently proposed another new approach called "Nnaturalness," said, "There are many theorists, myself included, who feel that we're in a totally unique time, where the questions on the table are the really huge, structural ones, not the details of the next particle. We're very lucky to get to live in a period like this— even if there may not be major, verified progress in our lifetimes."[3]

As theorists returned to their blackboards, the 6,000 experimentalists with CMS and ATLAS reveled in their exploration of a previously uncharted realm. "Nightmare, what does it mean?" said Spiropulu, referring to theorists' angst about the nightmare scenario. "We are exploring nature. Maybe we don't have time to think about nightmares like that, because we are being flooded in data and we are extremely excited."

There's still hope that new physics will show up. But discovering nothing, in Spiropulu's view, is a discovery all the same—especially when it heralds the death of cherished ideas. "Experimentalists have no religion," she said.

Some theorists agree. Talk of disappointment is "crazy talk," Arkani-Hamed said. "It's actually nature! We're learning the answer! These 6,000 people are busting their butts and you're pouting like a little kid because you didn't get the lollipop you wanted?"

TO SOLVE THE BIGGEST MYSTERY IN PHYSICS, JOIN TWO KINDS OF LAW

Robbert Dijkgraaf

S uppose aliens land on our planet and want to learn our current scientific knowledge. I would start with the 40-year-old documentary *Powers of Ten*. Granted, it's a bit out of date, but this short film, written and directed by the famous designer couple Charles and Ray Eames, captures in less than 10 minutes a comprehensive view of the cosmos.

The script is simple and elegant. When the film begins, we see a couple picnicking in a Chicago park. Then the camera zooms out. Every 10 seconds the field of vision gains a power of 10—from 10 meters across, to 100, to 1,000 and onward. Slowly the big picture reveals itself to us. We see the city, the continent, Earth, the solar system, neighboring stars, the Milky Way, all the way to the largest structures of the universe. Then in the second half of the film, the camera zooms in and delves into the smallest structures, uncovering more and more microscopic details. We travel into a human hand and discover cells, the double helix of the DNA molecule, atoms, nuclei and finally the elementary quarks vibrating inside a proton.

The movie captures the astonishing beauty of the macrocosm and microcosm, and it provides the perfect cliffhanger endings for conveying the challenges of fundamental science. As our then-8-year-old son asked when he first saw it, "How does it continue?" Exactly! Comprehending the next sequence is the aim of scientists who are pushing the frontiers of our understanding of the largest and smallest structures of the universe. Finally, I could explain what Daddy does at work!

Powers of Ten also teaches us that, while we traverse the various scales of length, time and energy, we also travel through different realms of knowledge. Psychology studies human behavior, evolutionary biology examines ecosystems, astrophysics investigates planets and stars, and cosmology concentrates on the universe as a whole. Similarly, moving inward, we navigate the subjects of biology, biochemistry, and atomic, nuclear and particle physics. It is as if the scientific disciplines are formed in strata, like the geological layers on display in the Grand Canyon.

Moving from one layer to another, we see examples of emergence and reductionism, these two overarching organizing principles of modern science. Zooming out, we see new patterns "emerge" from the complex behavior of individual building blocks. Biochemical reactions give rise to sentient beings. Individual organisms gather into ecosystems. Hundreds of billions of stars come together to make majestic swirls of galaxies.

As we reverse and take a microscopic view, we see reductionism at work. Complicated patterns dissolve into underlying simple bits. Life reduces to the reactions among DNA, RNA, proteins and other organic molecules. The complexity of chemistry flattens into the elegant beauty of the quantum mechanical atom. And, finally, the Standard Model of particle physics captures all known components of matter and radiation in just four forces and 17 elementary particles.

Which of these two scientific principles, reductionism or emergence, is more powerful? Traditional particle physicists would argue for reductionism; condensed-matter physicists, who study complex materials, for emergence. As articulated by the Nobel laureate (and particle physicist) David Gross: Where in nature do you find beauty, and where do you find garbage?

Take a look at the complexity of reality around us. Traditionally, particle physicists explain nature using a handful of particles and their interactions. But condensed matter physicists ask: What about an everyday glass of water? Describing its surface ripples in terms of the motions of the roughly 10^{24} individual water molecules—let alone their elementary particles—would be foolish. Instead of the impenetrable complexities at small scales (the "garbage") faced by traditional particle physicists, condensed matter physicists use the emergent laws, the "beauty" of hydrodynamics and thermodynamics. In fact, when we take the number of molecules to infinity (the equivalent of maximal garbage from a reductionist point of view), these laws of nature become crisp mathematical statements.

While many scientists praise the phenomenally successful reductionist approach of the past centuries, John Wheeler, the influential Princeton University physicist whose work touched on topics from nuclear physics to black holes, expressed an interesting alternative. "Every law of physics, pushed to the extreme, will be found to be statistical and approximate, not mathematically perfect and precise," he said. Wheeler pointed out an important feature of emergent laws: Their approximate nature allows for a certain flexibility that can accommodate future evolution.

In many ways, thermodynamics is the gold standard of an emergent law, describing the collective behavior of a large number of particles, irrespective of many microscopic details. It captures an astonishingly wide class of phenomena in succinct mathematical formulas. The laws hold in great universality—indeed, they were discovered before the atomic basis of matter was even established. And there are no loopholes. For example, the second law of thermodynamics states that a system's entropy—a measure of the amount of hidden microscopic information—will always grow in time.

Modern physics provides a precise language to capture the way things scale: the so-called renormalization group. This mathematical formalism allows us to go systematically from the small to the large. The essential step is taking averages. For example, instead of looking at the behavior of individual atoms that make up matter, we can take little cubes, say 10 atoms wide on each side, and take these cubes as our new building blocks. One can then repeat this averaging procedure. It is as if for each physical system one makes an individual *Powers of Ten* movie.

Renormalization theory describes in detail how the properties of a physical system change if one increases the length scale on which the observations are made. A famous example is the electric charge of particles that can increase or decrease depending on quantum interactions. A sociological example is understanding the behavior of groups of various sizes starting from individual behavior. Is there wisdom in crowds, or do the masses behave less responsibly?

Most interesting are the two endpoints of the renormalization process: the infinite large and infinite small. Here things will typically simplify because either all details are washed away, or the environment disappears. We see something like this with the two cliffhanger endings in *Powers of Ten*. Both the largest and the smallest structures of the universe are astonishingly simple. It is here that we find the two "standard models" of particle physics and cosmology.

Remarkably, modern insights about the most formidable challenge in theoretical physics—the push to develop a quantum theory of gravity—employ both the reductionist and emergent perspectives. Traditional approaches to quantum gravity, such as perturbative string theory, try to find a fully consistent microscopic description of all particles and forces. Such a "final theory" necessarily includes a theory of gravitons, the elementary particles of the gravitational field. For example, in string theory, the graviton is formed from a string that vibrates in a particular way. One of the early successes of string theory was a scheme to compute the behavior of such gravitons.

However, this is only a partial answer. Einstein taught us that gravity has a much wider scope: It addresses the structure of space and time. In a quantum-mechanical description, space and time would lose their meaning at ultrashort distances and time scales, raising the question of what replaces those fundamental concepts.

A complementary approach to combining gravity and quantum theory started with the groundbreaking ideas of Jacob Bekenstein and Stephen Hawking on the information content of black holes in the 1970s, and came into being with the seminal work of Juan Maldacena in the late 1990s. In this formulation, quantum space-time, including all the particles and forces in it, emerges from a completely different "holographic" description. The holographic system is quantum mechanical, but doesn't have any explicit form of gravity in it. Furthermore, it typically has fewer spatial dimensions. The system is, however, governed by a number that measures how large the system is. If one increases that number, the approximation to a classical gravitational system becomes more precise. In the end, space and time, together with Einstein's equations of general relativity, emerge out of the holographic system. The process is akin to the way that the laws of thermodynamics emerge out of the motions of individual molecules.

In some sense, this exercise is exactly the opposite of what Einstein tried to achieve. His aim was to build all of the laws of nature out of the dynamics of space and time, reducing physics to pure geometry. For him, space-time was the natural "ground level" in the infinite hierarchy of scientific objects—the bottom of the Grand Canyon. The present point of view thinks of space-time not as a starting point, but as an end point, as a natural structure that emerges out of the complexity of quantum information, much like the thermodynamics that rules our glass of water. Perhaps, in retrospect, it was not an accident that the two physical laws that Einstein liked best, thermodynamics and general relativity, have a common origin as emergent phenomena.

In some ways, this surprising marriage of emergence and reductionism allows one to enjoy the best of both worlds. For physicists, beauty is found at both ends of the spectrum.

THE STRANGE SECOND LIFE OF STRING THEORY

K. C. Cole

S tring theory strutted onto the scene some 30 years ago as perfection itself, a promise of elegant simplicity that would solve knotty problems in fundamental physics—including the notoriously intractable mismatch between Einstein's smoothly warped space-time and the inherently jittery, quantized bits of stuff that made up everything in it.

It seemed, to paraphrase Michael Faraday, much too wonderful *not* to be true: Simply replace infinitely small particles with tiny (but finite) vibrating loops of string. The vibrations would sing out quarks, electrons, gluons and photons, as well as their extended families, producing in harmony every ingredient needed to cook up the knowable world. Avoiding the infinitely small meant avoiding a variety of catastrophes. For one, quantum uncertainty couldn't rip space-time to shreds. At last, it seemed, here was a workable theory of quantum gravity.

Even more beautiful than the story told in words was the elegance of the math behind it, which had the power to make some physicists ecstatic.

To be sure, the theory came with unsettling implications. The strings were too small to be probed by experiment and lived in as many as 11 dimensions of space. These dimensions were folded in on themselves—or "compactified"—into complex origami shapes. No one knew just how the dimensions were compactified—the possibilities for doing so appeared to be endless—but surely some configuration would turn out to be just what was needed to produce familiar forces and particles.

For a time, many physicists believed that string theory would yield a unique way to combine quantum mechanics and gravity. "There was a hope. A moment," said David Gross, an original player in the so-called Princeton String Quartet, a Nobel Prize winner and permanent member of the Kavli Institute for Theoretical Physics at the University of California, Santa Barbara. "We even thought for a while in the mid-'80s that it was a unique theory."

And then physicists began to realize that the dream of one singular theory was an illusion. The complexities of string theory, all the possible permutations, refused to reduce to a single one that described our world. "After a certain point in the early '90s, people gave up on trying to connect to the real world," Gross said. "The last 20 years have really been a great extension of theoretical tools, but very little progress on understanding what's actually out there."

Many, in retrospect, realized they had raised the bar too high. Coming off the momentum of completing the solid and powerful Standard Model of particle physics in the 1970s, they hoped the story would repeat—only this time on a mammoth, all-embracing scale. "We've been trying to aim for the successes of the past where we had a very simple equation that captured everything," said Robbert Dijkgraaf, the director of the Institute for Advanced Study in Princeton, New Jersey. "But now we have this big mess."

Like many a maturing beauty, string theory has gotten rich in relationships, complicated, hard to handle and widely influential. Its tentacles have reached so deeply into so many areas in theoretical physics, it's become almost unrecognizable, even to string theorists. "Things have gotten almost postmodern," said Dijkgraaf, who is a painter as well as mathematical physicist.

The mathematics that have come out of string theory have been put to use in fields such as cosmology and condensed matter physics—the study of materials and their properties. It's so ubiquitous that "even if you shut down all the string theory groups, people in condensed matter, people in cosmology, people in quantum gravity will do it," Dijkgraaf said.

"It's hard to say really where you should draw the boundary around and say: This is string theory; this is not string theory," said Douglas Stanford, a physicist at the IAS. "Nobody knows whether to say they're a string theorist anymore," said Chris Beem, a mathematical physicist at the University of Oxford. "It's become very confusing."

String theory today looks almost fractal. The more closely people explore any one corner, the more structure they find. Some dig deep into particular crevices; others zoom out to try to make sense of grander patterns. The upshot is that string theory today includes much that no longer seems stringy. Those tiny loops of string whose harmonics were thought to breathe form into every particle and force known to nature (including elusive gravity) hardly even appear anymore on chalkboards at conferences. At the big annual string theory meeting in 2015, the Stanford University string theorist Eva Silverstein was amused to find she was one of the few giving a talk "on string theory proper," she said. A lot of the time she works on questions related to cosmology.

Even as string theory's mathematical tools get adopted across the physical sciences, physicists have been struggling with how to deal with the central tension of string theory: Can it ever live up to its initial promise? Could it ever give researchers insight into how gravity and quantum mechanics might be reconciled—not in a toy universe, but in our own?

"The problem is that string theory exists in the landscape of theoretical physics," said Juan Maldacena, perhaps the most prominent figure in the field today. "But we still don't know yet how it connects to nature as a theory of gravity." Maldacena now acknowledges the breadth of string theory, and its importance to many fields of physics—even those that don't require "strings" to be the fundamental stuff of the universe—when he defines string theory as "Solid Theoretical Research in Natural Geometric Structures."

AN EXPLOSION OF QUANTUM FIELDS

One high point for string theory as a theory of everything came in the late 1990s, when Maldacena revealed that a string theory including gravity in five dimensions was equivalent to a quantum field theory in four dimensions. This "AdS/CFT" duality appeared to provide a map for getting a handle on gravity—the most intransigent piece of the puzzle—by relating it to good old well-understood quantum field theory.

This correspondence was never thought to be a perfect real-world model. The five-dimensional space in which it works has an "anti-de Sitter" geometry, a strange M. C. Escher-ish landscape that is not remotely like our universe.

But researchers were surprised when they dug deep into the other side of the duality. Most people took for granted that quantum field theories— "bread and butter physics," Dijkgraaf calls them—were well understood and had been for half a century. As it turned out, Dijkgraaf said, "we only understand them in a very limited way."

These quantum field theories were developed in the 1950s to unify special relativity and quantum mechanics. They worked well enough for long enough that it didn't much matter that they broke down at very small scales and high energies. But today, when physicists revisit "the part you thought you understood 60 years ago," said Nima Arkani-Hamed, a physicist at the IAS, you find "stunning structures" that came as a complete surprise. "Every aspect of the idea that we understood quantum field theory turns out to be wrong. It's a vastly bigger beast."

Researchers have developed a huge number of quantum field theories in the past decade or so, each used to study different physical systems. Beem suspects there are quantum field theories that can't be described even in terms of quantum fields. "We have opinions that sound as crazy as that, in large part, because of string theory."

This virtual explosion of new kinds of quantum field theories is eerily reminiscent of physics in the 1930s, when the unexpected appearance of a new kind of particle—the muon—led a frustrated I. I. Rabi to ask: "Who ordered that?" The flood of new particles was so overwhelming by the 1950s that it led Enrico Fermi to grumble: "If I could remember the names of all these particles, I would have been a botanist."

Physicists began to see their way through the thicket of new particles only when they found the more fundamental building blocks making them up, like quarks and gluons. Now many physicists are attempting to do the same with quantum field theory. In their attempts to make sense of the zoo, many learn all they can about certain exotic species.

Conformal field theories (the right hand of AdS/CFT) are a starting point. You start with a simplified type of quantum field theory that behaves the same way at small and large distances, said David Simmons-Duffin, a physicist now at Caltech. If these specific kinds of field theories could be understood perfectly, answers to deep questions might become clear. "The idea is that if you understand the elephant's feet really, really well, you can interpolate in between and figure out what the whole thing looks like."

Like many of his colleagues, Simmons-Duffin says he's a string theorist mostly in the sense that it's become an umbrella term for anyone doing fundamental physics in underdeveloped corners. He's currently focusing on a physical system that's described by a conformal field theory but has nothing to do with strings. In fact, the system is water at its "critical point," where the distinction between gas and liquid disappears. It's interesting because water's behavior at the critical point is a complicated emergent system that arises from something simpler. As such, it could hint at dynamics behind the emergence of quantum field theories.

Beem focuses on supersymmetric field theories, another toy model, as physicists call these deliberate simplifications. "We're putting in some unrealistic features to make them easier to handle," he said. Specifically, they are amenable to tractable mathematics, which "makes it so a lot of things are calculable."

Toy models are standard tools in most kinds of research. But there's always the fear that what one learns from a simplified scenario does not apply to the real world. "It's a bit of a deal with the devil," Beem said. "String theory is a much less rigorously constructed set of ideas than quantum field theory, so you have to be willing to relax your standards a bit," he said. "But you're rewarded for that. It gives you a nice, bigger context in which to work."

It's the kind of work that makes people such as Sean Carroll of Caltech wonder if the field has strayed too far from its early ambitions—to find, if not a "theory of everything," at least a theory of quantum gravity. "Answering deep questions about quantum gravity has not really happened," he said. "They have all these hammers and they go looking for nails." That's fine, he said, even acknowledging that generations might be needed to develop a new theory of quantum gravity. "But it isn't fine if you forget that, ultimately, your goal is describing the real world."

It's a question he has asked his friends. Why are they investigating detailed quantum field theories? "What's the aspiration?" he asks. Their answers are logical, he says, but steps removed from developing a true description of our universe.

Instead, he's looking for a way to "find gravity inside quantum mechanics." A paper he wrote with colleagues claims to take steps toward just that.[1] It does not involve string theory.

THE BROAD POWER OF STRINGS

Perhaps the field that has gained the most from the flowering of string theory is mathematics itself. Sitting on a bench beside the IAS pond while watching a blue heron saunter in the reeds, Clay Córdova, a researcher there, explained how what seemed like intractable problems in mathematics were solved by imagining how the question might look to a string. For example, how many spheres could fit inside a Calabi-Yau manifold—the complex folded shape expected to describe how spacetime is compactified? Mathematicians had been stuck. But a two-dimensional string can wiggle around in such a complex space. As it wiggled, it could grasp new insights, like a mathematical multidimensional lasso. This was the kind of physical thinking Einstein was famous for: thought experiments about riding along with a light beam revealed $E = mc^2$. Imagining falling off a building led to his biggest eureka moment of all: Gravity is not a force; it's a property of space-time.

Using the physical intuition offered by strings, physicists produced a powerful formula for getting the answer to the embedded sphere question, and much more. "They got at these formulas using tools that mathematicians don't allow," Córdova said. Then, after string theorists found an answer, the mathematicians proved it on their own terms. "This is a kind of experiment," he explained. "It's an internal mathematical experiment." Not only was the stringy solution not wrong, it led to Fields Medal-winning mathematics. "This keeps happening," he said.

String theory has also made essential contributions to cosmology. The role that string theory has played in thinking about mechanisms behind the inflationary expansion of the universe—the moments immediately after the Big Bang, where quantum effects met gravity head on—is "surprisingly strong," said Silverstein, even though no strings are attached.

Still, Silverstein and colleagues have used string theory to discover, among other things, ways to see potentially observable signatures of various inflationary ideas. The same insights could have been found using quantum field theory, she said, but they weren't. "It's much more natural in string theory, with its extra structure."

Inflationary models get tangled in string theory in multiple ways, not least of which is the multiverse—the idea that ours is one of a perhaps infinite number of universes, each created by the same mechanism that begat our own. Between string theory and cosmology, the idea of an infinite landscape of possible universes became not just acceptable, but even taken for granted by a large number of physicists. The selection effect, Silverstein said, would be one quite natural explanation for why our world is the way it is: In a very different universe, we wouldn't be here to tell the story.

This effect could be one answer to a big problem string theory was supposed to solve. As Gross put it: "What picks out this particular theory"—the Standard Model—from the "plethora of infinite possibilities?"

Silverstein thinks the selection effect is actually a good argument for string theory. The infinite landscape of possible universes can be directly linked to "the rich structure that we find in string theory," she said—the innumerable ways that string theory's multidimensional space-time can be folded in upon itself.

BUILDING THE NEW ATLAS

At the very least, the mature version of string theory—with its mathematical tools that let researchers view problems in new ways—has provided powerful new methods for seeing how seemingly incompatible descriptions of nature can both be true. The discovery of dual descriptions of the same phenomenon pretty much sums up the history of physics. A century and a half ago, James Clerk Maxwell saw that electricity and magnetism were two sides of a coin. Quantum theory revealed the connection between particles and waves. Now physicists have strings.

"Once the elementary things we're probing spaces with are strings instead of particles," said Beem, the strings "see things differently." If it's too hard to get from A to B using quantum field theory, reimagine the problem in string theory, and "there's a path," Beem said.

In cosmology, string theory "packages physical models in a way that's easier to think about," Silverstein said. It may take centuries to tie together all these loose strings to weave a coherent picture, but young researchers like Beem aren't bothered a bit. His generation never thought string theory was going to solve everything. "We're not stuck," he said. "It doesn't feel like we're on the verge of getting it all sorted, but I know more each day than I did the day before—and so presumably we're getting somewhere."

Stanford thinks of it as a big crossword puzzle. "It's not finished, but as you start solving, you can tell that it's a valid puzzle," he said. "It's passing consistency checks all the time."

"Maybe it's not even possible to capture the universe in one easily defined, self-contained form, like a globe," Dijkgraaf said, sitting in Robert Oppenheimer's many windowed office from when he was Einstein's boss, looking over the vast lawn at the IAS, the pond and the woods in the distance. Einstein, too, tried and failed to find a theory of everything, and it takes nothing away from his genius.

"Perhaps the true picture is more like the maps in an atlas, each offering very different kinds of information, each spotty," Dijkgraaf said. "Using the atlas will require that physics be fluent in many languages, many approaches, all at the same time. Their work will come from many different directions, perhaps far-flung."

He finds it "totally disorienting" and also "fantastic."

Arkani-Hamed believes we are in the most exciting epoch of physics since quantum mechanics appeared in the 1920s. But nothing will happen quickly. "If you're excited about responsibly attacking the very biggest existential physics questions ever, then you should be excited," he said. "But if you want a ticket to Stockholm for sure in the next 15 years, then probably not."

A FIGHT FOR THE SOUL OF SCIENCE

Natalie Wolchover

P hysicists typically think they "need philosophers and historians of science like birds need ornithologists," the Nobel laureate David Gross told a roomful of philosophers, historians and physicists in Munich, Germany, paraphrasing Richard Feynman.

But desperate times call for desperate measures.

Fundamental physics faces a problem, Gross explained—one dire enough to call for outsiders' perspectives. "I'm not sure that we don't need each other at this point in time," he said.

It was the opening session of a three-day workshop, held in a Romanesque-style lecture hall at Ludwig Maximilian University (LMU Munich) in December 2015, one year after George Ellis and Joe Silk, two white-haired physicists now sitting in the front row, called for such a conference in an incendiary opinion piece in *Nature*.[1] One hundred attendees had descended on a land with a celebrated tradition in both physics and the philosophy of science to wage what Ellis and Silk declared a "battle for the heart and soul of physics."

The crisis, as Ellis and Silk tell it, is the wildly speculative nature of modern physics theories, which they say reflects a dangerous departure from the scientific method. Many of today's theorists—chief among them the proponents of string theory and the multiverse hypothesis—appear convinced of their ideas on the grounds that they are beautiful or logically compelling, despite the impossibility of testing them. Ellis and Silk accused these theorists of "moving the goalposts" of science and blurring the line between physics and pseudoscience. "The imprimatur of science should be awarded only to a theory that is testable," Ellis and Silk wrote, thereby disqualifying most of the leading theories of the past 40 years. "Only then can we defend science from attack."

They were reacting, in part, to the controversial ideas of Richard Dawid, an Austrian philosopher whose 2013 book *String Theory and the Scientific Method* identified three kinds of "nonempirical" evidence that Dawid says can help build trust in scientific theories absent empirical data. Dawid, then a researcher at LMU Munich, answered Ellis and Silk's battle cry and assembled far-flung scholars anchoring all sides of the argument for the high-profile event.

Gross, a supporter of string theory who won the 2004 Nobel Prize in physics for his work on the force that glues atoms together, kicked off the workshop by asserting that the problem lies not with physicists but with a "fact of nature"—one that we have been approaching inevitably for four centuries.

The dogged pursuit of a fundamental theory governing all forces of nature requires physicists to inspect the universe more and more closely—to examine, for instance, the atoms within matter, the protons and neutrons within those atoms, and the quarks within those protons and neutrons. But this zooming in demands ever more energy, and the difficulty and cost of building new machines increases exponentially relative to the energy requirement, Gross said. "It hasn't been a problem so much for the last 400 years, where we've gone from centimeters to millionths of a millionth of a millionth of a centimeter"—the current resolving power of the Large Hadron Collider (LHC) in Switzerland, he said. "We've gone very far, but this energy-squared is killing us."

As we approach the practical limits of our ability to probe nature's underlying principles, the minds of theorists have wandered far beyond the tiniest observable distances and highest possible energies. Strong clues indicate that the truly fundamental constituents of the universe lie at a distance scale 10 million billion times smaller than the resolving power of the LHC. This is the domain of nature that string theory, a candidate "theory of everything," attempts to describe. But it's a domain that no one has the faintest idea how to access.

The problem also hampers physicists' quest to understand the universe on a cosmic scale: No telescope will ever manage to peer past our universe's cosmic horizon and glimpse the other universes posited by the multiverse hypothesis. Yet modern theories of cosmology lead logically to the possibility that our universe is just one of many.

Whether the fault lies with theorists for getting carried away, or with nature, for burying its best secrets, the conclusion is the same: Theory has detached itself from experiment. The objects of theoretical speculation are now too far away, too small, too energetic or too far in the past to reach or rule out with our earthly instruments. So, what is to be done? As Ellis and Silk wrote, "Physicists, philosophers and other scientists should hammer out a new narrative for the scientific method that can deal with the scope of modern physics."

The Ends of Evidence

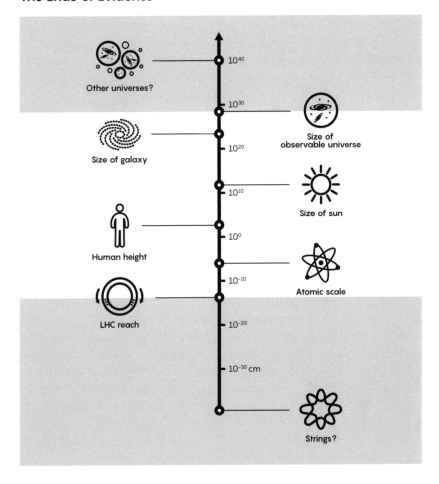

FIGURE 8.2
Humans can probe the universe over a vast range of scales (white area), but many modern physics theories involve scales outside of this range (gray). Icons via Freepik.

"The issue in confronting the next step," said Gross, "is not one of ideology but strategy: What is the most useful way of doing science?"

Over those three mild winter days, scholars grappled with the meaning of *theory*, *confirmation* and *truth*; how science works; and whether, in this day and age, philosophy should guide research in physics or the other way around. Over the course of these pressing yet timeless discussions, a degree of consensus took shape.

RULES OF THE GAME

Throughout history, the rules of science have been written on the fly, only to be revised to fit evolving circumstances. The ancients believed they could reason their way toward scientific truth. Then, in the 17th century, Isaac Newton ignited modern science by breaking with this "rationalist" philosophy, adopting instead the "empiricist" view that scientific knowledge derives only from empirical observation. In other words, a theory must be proved experimentally to enter the book of knowledge.

But what requirements must an untested theory meet to be considered scientific? Theorists guide the scientific enterprise by dreaming up the ideas to be put to the test and then interpreting the experimental results; what keeps theorists within the bounds of science?

Today, most physicists judge the soundness of a theory by using the Austrian-British philosopher Karl Popper's rule of thumb. In the 1930s, Popper drew a line between science and nonscience in comparing the work of Albert Einstein with that of Sigmund Freud. Einstein's theory of general relativity, which cast the force of gravity as curves in space and time, made risky predictions—ones that, if they hadn't succeeded so brilliantly, would have failed miserably, falsifying the theory. But Freudian psychoanalysis was slippery: Any fault of your mother's could be worked into your diagnosis. The theory wasn't falsifiable, and so, Popper decided, it wasn't science.

Critics accuse string theory and the multiverse hypothesis, as well as cosmic inflation—the leading theory of how the universe began—of falling on the wrong side of Popper's line of demarcation. To borrow the title of the Columbia University physicist Peter Woit's 2006 book on string theory, these ideas are "not even wrong," say critics. In their editorial, Ellis and Silk invoked the spirit of Popper: "A theory must be falsifiable to be scientific."

But, as many in Munich were surprised to learn, falsificationism is no longer the reigning philosophy of science. Massimo Pigliucci, a philosopher at the Graduate Center of the City University of New York, pointed out that falsifiability is woefully inadequate as a separator of science and nonscience, as Popper himself recognized. Astrology, for instance, is falsifiable—indeed, it has been falsified *ad nauseam*—and yet it isn't science. Physicists' preoccupation with Popper "is really something that needs to stop," Pigliucci said. "We need to talk about current philosophy of science. We don't talk about something that was current 50 years ago."

Nowadays, as several philosophers at the workshop said, Popperian falsificationism has been supplanted by Bayesian confirmation theory, or Bayesianism, a modern framework based on the 18th-century probability theory of the English statistician and minister Thomas Bayes. Bayesianism allows for the fact that modern scientific theories typically make claims far beyond what can be directly observed—no one has ever seen an atom—and so today's theories often resist a falsified-unfalsified dichotomy. Instead, trust in a theory often falls somewhere along a continuum, sliding up or down between 0 and 100 percent as new information becomes available. "The Bayesian framework is much more flexible" than Popper's theory, said Stephan Hartmann, a Bayesian philosopher at LMU. "It also connects nicely to the psychology of reasoning."

Gross concurred, saying that, upon learning about Bayesian confirmation theory from Dawid's book, he felt "somewhat like the Molière character who said, 'Oh my God, I've been talking prose all my life!'"

Another advantage of Bayesianism, Hartmann said, is that it is enabling philosophers like Dawid to figure out "how this nonempirical evidence fits in, or can be fit in."

ANOTHER KIND OF EVIDENCE

Dawid, who is mild-mannered and smiley with floppy brown hair, started his career as a theoretical physicist. In the late 1990s, during a stint at the University of California, Berkeley, a hub of string-theory research, Dawid became fascinated by how confident many string theorists seemed to be that they were on the right track, despite string theory's complete lack of empirical support. "Why do they trust the theory?" he recalls wondering. "Do they have different ways of thinking about it than the canonical understanding?"

String theory says that elementary particles have dimensionality when viewed close-up, appearing as wiggling loops (or "strings") and membranes at nature's highest zoom level. According to the theory, extra dimensions also materialize in the fabric of space itself. The different vibrational modes of the strings in this higher-dimensional space give rise to the spectrum of particles that make up the observable world. In particular, one of the vibrational modes fits the profile of the "graviton"—the hypothetical particle associated with the force of gravity. Thus, string theory unifies gravity, now described by Einstein's theory of general relativity, with the rest of particle physics.

However string theory, which has its roots in ideas developed in the late 1960s, has made no testable predictions about the observable universe. To understand why so many researchers trust it anyway, Dawid signed up for some classes in philosophy of science, and upon discovering how little study had been devoted to the phenomenon, he switched fields.

In the early 2000s, he identified three nonempirical arguments that generate trust in string theory among its proponents. First, there appears to be only one version of string theory capable of achieving unification in a consistent way (though it has many different mathematical representations); furthermore, no other "theory of everything" capable of unifying all the fundamental forces has been found, despite immense effort. (A rival approach called loop quantum gravity describes gravity at the quantum scale, but makes no attempt to unify it with the other forces.) This "no-alternatives" argument, colloquially known as "string theory is the only game in town," boosts theorists' confidence that few or no other possible unifications of the four fundamental forces exist, making it more likely that string theory is the right approach.

Second, string theory grew out of the Standard Model—the accepted, empirically validated theory incorporating all known fundamental particles and forces (apart from gravity) in a single mathematical structure—and the Standard Model also had no alternatives during its formative years. This "meta-inductive" argument, as Dawid calls it, buttresses the no-alternatives argument by showing that it has worked before in similar contexts, countering the possibility that physicists simply aren't clever enough to find the alternatives that exist.

The third nonempirical argument is that string theory has unexpectedly delivered explanations for several other theoretical problems aside from the unification problem it was intended to address. The staunch string theorist Joseph Polchinski of the University of California, Santa Barbara, presented several examples of these "unexpected explanatory interconnections," as Dawid has termed them, in a paper read in Munich in his absence.[2] String theory explains the entropy of black holes, for example, and, in a surprising discovery that has caused a surge of research in the past 15 years, is mathematically translatable into a theory of particles, such as the theory describing the nuclei of atoms.

Polchinski concluded that, considering how far away we are from the exceptionally fine grain of nature's fundamental distance scale, we should count ourselves lucky: "String theory exists, and we have found it." (Polchinski also used Dawid's nonempirical arguments to calculate the Bayesian odds that the multiverse exists as 94 percent—a value that has been ridiculed by the Internet's vocal multiverse critics.)

One concern with including nonempirical arguments in Bayesian confirmation theory, Dawid acknowledged in his talk, is "that it opens the floodgates to abandoning all scientific principles." One can come up with all kinds of nonempirical virtues when arguing in favor of a pet idea. "Clearly the risk is there, and clearly one has to be careful about this kind of reasoning," Dawid said. "But acknowledging that nonempirical confirmation is part of science, and has been part of science for quite some time, provides a better basis for having that discussion than pretending that it wasn't there, and only implicitly using it, and then saying I haven't done it. Once it's out in the open, one can discuss the pros and cons of those arguments within a specific context."

THE MUNICH DEBATE

The trash heap of history is littered with beautiful theories. The Danish historian of cosmology Helge Kragh, who detailed a number of these failures in his 2011 book, *Higher Speculations*, spoke in Munich about the 19th-century vortex theory of atoms. This "Victorian theory of everything," developed by the Scots Peter Tait and Lord Kelvin, postulated that atoms are microscopic vortexes in the ether, the fluid medium that was believed at the time to fill space. Hydrogen, oxygen and all other atoms were, deep down, just different types of vortical knots. At first, the theory "seemed to be highly promising," Kragh said. "People were fascinated by the richness of the mathematics, which could keep mathematicians busy for centuries, as was said at the time." Alas, atoms are not vortexes, the ether does not exist, and theoretical beauty is not always truth.

Except sometimes it is. Rationalism guided Einstein toward his theory of relativity, which he believed in wholeheartedly on rational grounds before it was ever tested. "I hold it true that pure thought can grasp reality, as the ancients dreamed," Einstein said in 1933, years after his theory had been confirmed by observations of starlight bending around the sun.

The question for the philosophers is: Without experiments, is there any way to distinguish between the nonempirical virtues of vortex theory and those of Einstein's theory? Can we ever really trust a theory on nonempirical grounds?

In discussions on the third afternoon of the workshop, the philosopher Radin Dardashti asserted that Dawid's philosophy specifically aims to pinpoint which nonempirical arguments should carry weight, allowing scientists to "make an assessment that is not based on simplicity, which is not based on beauty." Dawidian assessment is meant to be more objective than these measures, Dardashti explained—and more revealing of a theory's true promise.

Gross said Dawid has "described beautifully" the strategies physicists use "to gain confidence in a speculation, a new idea, a new theory."

"You mean confidence that it's true?" asked Peter Achinstein, an 80-year-old philosopher and historian of science at Johns Hopkins University. "Confidence that it's useful? Confidence that ..."

"Let's give an operational definition of confidence: I will continue to work on it," Gross said.

"That's pretty low," Achinstein said.

"Not for science," Gross said. "That's the question that matters."

Kragh pointed out that even Popper saw value in the kind of thinking that motivates string theorists today. Popper called speculation that did not yield testable predictions "metaphysics," but he considered such activity worthwhile, since it might become testable in the future. This was true of atomic theory, which many 19th-century physicists feared would never be empirically confirmed. "Popper was not a naive Popperian," Kragh said. "If a theory is not falsifiable," Kragh said, channeling Popper, "it should not be given up. We have to wait."

But several workshop participants raised qualms about Bayesian confirmation theory, and about Dawid's nonempirical arguments in particular.

Carlo Rovelli, a proponent of loop quantum gravity (string theory's rival) who is based at Aix-Marseille University in France, objected that Bayesian confirmation theory does not allow for an important distinction that exists in science between theories that scientists are certain about and those that are still being tested. The Bayesian "confirmation" that atoms exist is essentially 100 percent, as a result of countless experiments. But Rovelli says that the degree of confirmation of atomic theory shouldn't even be measured in the same units as that of string theory. String theory is not, say, 10 percent as confirmed as atomic theory; the two have different statuses entirely. "The problem with Dawid's 'nonempirical confirmation' is that it muddles the point," Rovelli said. "And of course some string theorists are happy of muddling it this way, because they can then say that string theory is 'confirmed,' equivocating."

The German physicist Sabine Hossenfelder, in her talk, argued that progress in fundamental physics very often comes from abandoning cherished prejudices (such as, perhaps, the assumption that the forces of nature must be unified). Echoing this point, Rovelli said "Dawid's idea of nonempirical confirmation [forms] an obstacle to this possibility of progress, because it bases our credence on our own previous credences." It "takes away one of the tools—maybe the soul itself—of scientific thinking," he continued, "which is 'do not trust your own thinking.'"

One important outcome of the workshop, according to Ellis, was an acknowledgment by participating string theorists that the theory is not "confirmed" in the sense of being verified. "David Gross made his position clear: Dawid's criteria are good for justifying working on the theory, not for saying the theory is validated in a nonempirical way," Ellis wrote in an email. "That seems to me a good position—and explicitly stating that is progress."

In considering how theorists should proceed, many attendees expressed the view that work on string theory and other as-yet-untestable ideas should continue. "Keep speculating," Achinstein wrote in an email after the workshop, but "give your motivation for speculating, give your explanations, but admit that they are only possible explanations."

"Maybe someday things will change," Achinstein added, "and the speculations will become testable; and maybe not, maybe never." We may never know for sure the way the universe works at all distances and all times, "but perhaps you can narrow the live possibilities to just a few," he said. "I think that would be some progress."

ACKNOWLEDGMENTS

A publication is only as good as the people who produce it—people plural, because little of value in journalism or publishing is achieved as a solitary act. First, I give my heartfelt thanks to the many tremendously talented writers, editors and artists who crafted the words and pictures in this book, breathing life into these wonderfully illuminating science stories. I especially want to acknowledge senior physics writer Natalie Wolchover and former senior biology writer Emily Singer for their many contributions to this collection.

In addition to the bylined authors, I wish to thank my esteemed magazine coeditors, Michael Moyer and John Rennie, for graciously and intelligently reviewing story ideas, assigning articles, guiding writers and safeguarding *Quanta*'s standards; art director Olena Shmahalo for her sublime vision that permeates the magazine's visual identity; graphics editor Lucy Reading-Ikkanda for transforming impossibly abstract concepts into elegant, accessible visualizations; contributing artist Sherry Choi for retooling the graphics for the book in a beautiful and consistent style; artist Filip Hodas for creating the imaginative covers; unsung heroes like Roberta Klarreich and all of our contributing copy editors for cleaning and polishing our prose, and Matt Mahoney and all of our contributing fact checkers for acting as a last line of defense and allowing me to sleep better at night; Molly Frances for her meticulous formatting of the reference notes; and producers Jeanette Kazmierczak and Michelle Yun for doing the little things without which everything would come to a grinding halt.

Quanta Magazine, and these books by extension, would not exist without the generous support of the Simons Foundation. I would like to express my deepest gratitude to foundation leaders Jim and Marilyn Simons and Marion Greenup for believing in this project and nurturing it with kindness, wisdom and intellectual rigor every step of the way. I thank Stacey Greenebaum for her creative public outreach efforts, Jennifer Maimone-Medwick and

Yolaine Seaton for their conscientious reviewing of contract language, the entire *Quanta* team for being the best in the business, our distinguished advisory board members for their invaluable counsel and our wonderful foundation colleagues—too many to name individually—for making our work lives both easier and more entertaining.

Special thanks to foreword writer Sean Carroll, who took precious time out of his whirlwind schedule as a Caltech theoretical physicist, public speaker and acclaimed science writer to read an early draft of this volume and share his expert perspective.

I'd like to thank the amazing team at the MIT Press, starting with acquisitions editor Jermey Matthews and book designer Yasuyo Iguchi, who have been a delight to work with, and with a special shout-out to director Amy Brand, who first reached out to me about publishing a *Quanta* book and who provided the leadership, enthusiasm and resources that allowed this project to thrive.

Producing a book is like assembling a big Rube Goldberg machine with countless moving parts and endless opportunities for mistakes and failure. I'm lucky to have found book agent Jeff Shreve of the Science Factory, whose wise feedback and guidance played no small part in averting any number of missteps and making this book a reality.

I'm grateful to the scientists and mathematicians who answered calls from our reporters, editors and fact checkers and patiently and sure-footedly guided us through treacherous territory filled with technical land mines.

I owe everything to my parents, David and Lydia, who gifted me with a lifelong appreciation for science and math; my brother, Ben, who is an inspiration as a high school math teacher; and my wife, Genie, and sons, Julian and Tobias, who give life infinite meaning.

—Thomas Lin

CONTRIBUTORS

Philip Ball is a science writer and author based in London who contributes to *Quanta Magazine, Nature, New Scientist, Prospect, Nautilus* and *The Atlantic*, among other publications. His books include *The Water Kingdom, Bright Earth, Invisible* and most recently, *Beyond Weird*.

K. C. Cole is the author of eight nonfiction books, including *The Universe and the Teacup: The Mathematics of Truth and Beauty* and *Something Incredibly Wonderful Happens*. She is a former science correspondent for *The Los Angeles Times* and her work has also appeared in *The New Yorker, The New York Times* and *Discover*.

Robbert Dijkgraaf is director and Leon Levy Professor at the Institute for Advanced Study in Princeton, New Jersey. He is an author, with Abraham Flexner, of *The Usefulness of Useless Knowledge*.

Dan Falk is a science journalist based in Toronto, Canada. His books include *In Search of Time* and *The Science of Shakespeare*.

Courtney Humphries is a freelance writer based in Boston who covers science, nature, medicine and the built environment. Her work has appeared in *Quanta Magazine*, the *Boston Globe, Nautilus, Nature, Technology Review* and *CityLab*, among other publications. She was recently a Knight Science Journalism Fellow at the Massachusetts Institute of Technology.

Ferris Jabr is a writer based in Portland, Oregon. He has written for *Quanta Magazine, Scientific American, The New York Times Magazine, Outside* and *Lapham's Quarterly*, among other publications.

Thomas Lin is the founding editor in chief of *Quanta Magazine*. He previously managed the online science and national news sections at the *New York Times*, where he won a White House News Photographers Association's

"Eyes of History" award, and wrote about science, tennis and technology. He has also written for the *New Yorker, Tennis Magazine* and other publications.

Katia Moskvitch is a science and technology journalist based in London. She has written about physics, astronomy and other topics for *Quanta Magazine, Nature, Science, Scientific American, The Economist, Nautilus, New Scientist*, the *BBC* and other publications. A mechanical engineer by training, she is the author of the book *Call me 'Pops': Le Bon Dieu Dans La Rue*.

George Musser is a contributing editor at *Scientific American* and the author of two books, *Spooky Action at a Distance* and *The Complete Idiot's Guide to String Theory*. He is the recipient of the 2011 American Institute of Physics Science Writing Award and was a Knight Science Journalism Fellow at MIT from 2014 to 2015.

Michael Nielsen is a research fellow at YC Research in San Francisco. He has written books on quantum computing, open science and deep learning. His current research interests are collective cognition and intelligence amplification.

Jennifer Ouellette is a freelance writer and an author of popular science books, including *Me, Myself, and Why: Searching for the Science of Self*. Her work has also appeared in *Quanta Magazine, Discover, New Scientist, Smithsonian, Nature, Physics Today, Physics World, Slate* and *Salon*.

John Pavlus is a writer and filmmaker whose work has appeared in *Quanta Magazine, Scientific American, Bloomberg Businessweek* and *The Best American Science and Nature Writing* series. He lives in Portland, Oregon.

Emily Singer is a former senior biology writer and contributing editor at *Quanta Magazine*. Previously, she was the news editor for *SFARI.org* and the biomedical editor for *Technology Review*. She has written for *Nature, New Scientist*, the *Los Angeles Times* and the *Boston Globe*, and has a master's in neuroscience from the University of California, San Diego.

Andreas von Bubnoff is an award-winning science journalist and multimedia producer based in New York City. His work has appeared in *Quanta Magazine, The Los Angeles Times*, the *Chicago Tribune, Nautilus, Nature, Die Zeit*, the *Frankfurter Allgemeine Zeitung* and *RiffReporter*.

Frank Wilczek was awarded the 2004 Nobel Prize in physics for his work on the theory of the strong force. His most recent book is *A Beautiful Question: Finding Nature's Deep Design*. Wilczek is the Herman Feshbach Professor of Physics at the Massachusetts Institute of Technology and a professor of physics at Stockholm University.

Natalie Wolchover is a senior writer at *Quanta Magazine* covering the physical sciences. Her work has been featured in *The Best Writing on Mathematics* and recognized with the 2016 Evert Clark/Seth Payne Award and the 2017 American Institute of Physics Science Writing Award. She studied graduate-level physics at the University of California, Berkeley.

Carl Zimmer is a science columnist for *The New York Times* and the author of 13 books, including *She Has Her Mother's Laugh* and *Parasite Rex*.

NOTES

Is Nature Unnatural?

1. Marco Farina, Duccio Pappadopulo, and Alessandro Strumia, "A Modified Naturalness Principle and Its Experimental Tests," *Journal of High Energy Physics* 2013, no. 22 (August 2013), https://arxiv.org/abs/1303.7244.

2. Steven Weinberg, "Anthropic Bound on the Cosmological Constant," *Physical Review Letters* 59, no. 22 (November 30, 1987), https://journals.aps.org/prl/abstract/10.1103/PhysRevLett.59.2607.

3. Raphael Bousso and Joseph Polchinski, "Quantization of Four-Form Fluxes and Dynamical Neutralization of the Cosmological Constant," *Journal of High Energy Physics* 2000, no. 6 (June 2000), https://arxiv.org/pdf/hep-th/0004134.pdf.

4. Raphael Bousso and Lawrence Hall, "Why Comparable? A Multiverse Explanation of the Dark Matter-Baryon Coincidence," *Physical Review D* 88, no. 6 (September 2013), https://arxiv.org/abs/1304.6407.

5. V. Agrawal et al., "The Anthropic Principle and the Mass Scale of the Standard Model," *Physical Review D* 57, no. 9 (May 1, 1998), https://arxiv.org/pdf/hep-ph/9707380v2.pdf.

Alice and Bob Meet the Wall of Fire

1. Ahmed Almheiri et al., "Black Holes: Complementarity or Firewalls?" *Journal of High Energy Physics* 2013, no. 62 (February 2013), https://arxiv.org/pdf/1207.3123.pdf.

2. Leonard Susskind, "Singularities, Firewalls, and Complementarity," *Journal of High Energy Physics* (August 16, 2012), https://arxiv.org/abs/1208.3445.

3. Raphael Bousso, "Complementarity Is Not Enough," *Physical Review D* 87, no. 12 (June 20, 2013), https://arxiv.org/abs/arXiv:1207.5192.

Wormholes Untangle a Black Hole Paradox

1. Almheiri, "Black Holes: Complementarity or Firewalls?," 62.

2. Juan Maldacena and Leonard Susskind, "Cool Horizons for Entangled Black Holes," *Fortschritte der Physik* 61, no. 9 (September 2013): 781–811, https://arxiv.org /abs/1306.0533.

3. Juan Maldacena, Stephen H. Shenker, and Douglas Stanford, "A Bound on Chaos," *Journal of High Energy Physics* 2016, no. 8 (August 2016), https://arxiv.org/abs/1503 .01409.

4. Vijay Balasubramanian et al., "Multiboundary Wormholes and Holographic Entanglement," *Classical and Quantum Gravity* 31, no. 18 (September 2014), https:// arxiv.org/abs/1406.2663.

How Quantum Pairs Stitch Space-Time

1. Román Orús, "Advances on Tensor Network Theory: Symmetries, Fermions, Entanglement, and Holography," *European Physical Journal B* 87, no. 11 (November 2014), https://arxiv.org/abs/1407.6552.

2. Shinsei Ryu and Tadashi Takayanagi, "Aspects of Holographic Entanglement Entropy," *Journal of High Energy Physics* 2006, no. 8 (August 2006), https://arxiv.org /abs/hep-th/0605073.

3. Brian Swingle and Mark Van Raamsdonk, "Universality of Gravity from Entanglement" (May 12, 2014), https://arxiv.org/abs/1405.2933.

In a Multiverse, What Are the Odds?

1. Alan H. Guth, "Inflationary Universe: A Possible Solution to the Horizon and Flatness Problems," *Physical Review D* 23, no. 2 (January 15, 1981), https://journals .aps.org/prd/abstract/10.1103/PhysRevD.23.347.

2. Weinberg, "Anthropic Bound on the Cosmological Constant."

3. Agrawal et al., "The Anthropic Principle and the Mass Scale of the Standard Model."

4. Abraham Loeb, "An Observational Test for the Anthropic Origin of the Cosmological Constant," *Journal of Cosmology and Astroparticle Physics* 2006 (April 2006), https://arxiv.org/abs/astro-ph/0604242.

5. Jaume Garriga and Alexander Vilenkin, "Watchers of the Multiverse," *Journal of Cosmology and Astroparticle Physics* 2013 (May 2013), https://arxiv.org/abs/1210.7540.

6. Alan H. Guth and Vitaly Vanchurin, "Eternal Inflation, Global Time Cutoff Measures, and a Probability Paradox" (August 2, 2011), https://arxiv.org/abs/1108.0665.

7. Raphael Bousso, "Holographic Probabilities in Eternal Inflation," *Physical Review Letters* 97, no. 19 (November 2006), https://arxiv.org/abs/hep-th/0605263.

8. Raphael Bousso et al., "Predicting the Cosmological Constant from the Causal Entropic Principle," *Physical Review D* 76, no. 4 (August 2007), https://arxiv.org/abs/hep-th/0702115.

9. Bousso, "Why Comparable? A Multiverse Explanation."

Multiverse Collisions May Dot the Sky

1. Stephen M. Feeney et al., "First Observational Tests of Eternal Inflation," *Physical Review Letters* 107, no. 7 (August 2011), https://journals.aps.org/prl/abstract/10.1103/PhysRevLett.107.071301.

2. Stephen M. Feeney et al., "Forecasting Constraints from the Cosmic Microwave Background on Eternal Inflation," *Physical Review D* 92, no. 8 (October 16, 2015), https://arxiv.org/pdf/1506.01716.pdf.

3. John T. Giblin Jr., et al., "How to Run through Walls: Dynamics of Bubble and Soliton Collisions," *Physical Review D* 82, no. 4 (August 2010), https://journals.aps.org/prd/abstract/10.1103/PhysRevD.82.045019.

How Feynman Diagrams Almost Saved Space

1. Frank Wilczek, "Viewpoint: A Landmark Proof," *Physics* 4, no. 10 (February 2011), https://physics.aps.org/articles/v4/10.

2. David Lindley, "Focus: Landmarks–Lamb Shift Verifies New Quantum Concept," *Physics* 5, no. 83 (July 27, 2012), https://physics.aps.org/articles/v5/83.

3. R. P. Feynman, "Space-Time Approach to Quantum Electrodynamics," *Physical Review* 76, no. 6 (September 1949), https://doi.org/10.1103/PhysRev.76.769.

A Jewel at the Heart of Quantum Physics

1. Nima Arkani-Hamed and Jaroslav Trnka, "The Amplituhedron" (December 6, 2013), https://arxiv.org/pdf/1312.2007.pdf.

2. Nima Arkani-Hamed et al., "Scattering Amplitudes and the Positive Grassmannian" (March 17, 2014), https://arxiv.org/abs/1212.5605.

New Support for Alternative Quantum View

1. Berthold-George Englert et al., "Surrealistic Bohm Trajectories," *Zeitschrift für Naturforschung* A 47, no. 12: 1175–1186. https://doi.org/10.1515/zna-1992-1201.

2. Dylan H. Mahler et al., "Experimental Nonlocal and Surreal Bohmian Trajectories," *Science Advances* 2, no. 2 (February 2016), https://doi.org/10.1126/sciadv.1501466.

3. Englert et al., "Surrealistic Bohm Trajectories."

4. Mahler et al., "Experimental Nonlocal and Surreal Bohmian Trajectories."

Entanglement Made Simple

1. Daniel M. Greenberger, Michael A. Horne, and Anton Zeilinger, "Going Beyond Bell's Theorem," in *Bell's Theorem, Quantum Theory, and Conceptions of the Universe*, ed. M. Kafatos (Dordrecht: Kluwer Academic Publishers, 1989), 69–72, https://arxiv.org /abs/0712.0921.

Quantum Theory Rebuilt From Simple Physical Principles

1. Maximilian Schlosshauer, Johannes Kofler, and Anton Zeilinger, "A Snapshot of Foundational Attitudes toward Quantum Mechanics," *Studies in History and Philosophy of Science Part B: Studies in History and Philosophy of Modern Physics* 44, no. 3 (August 2013): 222–230, https://arxiv.org/abs/1301.1069.

2. Lucien Hardy, "Quantum Theory from Five Reasonable Axioms" (September 25, 2001), https://arxiv.org/abs/quant-ph/0101012.

3. Lucien Hardy, "Reformulating and Reconstructing Quantum Theory" (August 25, 2011), https://arxiv.org/abs/1104.2066.

4. G. Chiribella, G. M. D'Ariano, and P. Perinotti, "Informational Derivation of Quantum Theory," *Physical Review A* 84, no. 1 (July 2011), https://arxiv.org/abs/1011.6451.

5. Rob Clifton, Jeffrey Bub, and Hans Halvorson, "Characterizing Quantum Theory in Terms of Information-Theoretic Constraints," *Foundations of Physics* 33, no. 11 (November 2003), https://arxiv.org/abs/quant-ph/0211089.

6. Borivoje Dakic and Caslav Brukner, "Quantum Theory and Beyond: Is Entanglement Special?" (November 3, 2009), https://arxiv.org/abs/0911.0695.

7. Lucien Hardy, "Quantum Gravity Computers: On the Theory of Computation with Indefinite Causal Structure," in *Quantum Reality, Relativistic Causality, and Closing the Epistemic Circle*, The Western Ontario Series in Philosophy of Science, Vol. 73 (Dordrecht: Springer, 2009), 379–401, https://arxiv.org/abs/quant-ph/0701019.

8. Giulio Chiribella, "Perfect Discrimination of No-Signalling Channels via Quantum Superposition of Causal Structures," *Physical Review A* 86, no. 4 (October 2012), https://arxiv.org/abs/1109.5154.

Time's Arrow Traced to Quantum Source

1. Noah Linden et al., "Quantum Mechanical Evolution towards Thermal Equilibrium," *Physical Review E* 79, no. 6 (June 2009), https://doi.org/10.1103/PhysRevE.79.061103.

2. Peter Reimann, "Foundation of Statistical Mechanics under Experimentally Realistic Conditions," *Physical Review Letters* 101, no. 19 (November 7, 2008), https://doi.org/10.1103/PhysRevLett.101.190403.

3. Anthony J. Short and Terence C. Farrelly, "Quantum Equilibration in Finite Time," *New Journal of Physics* 14 (January 2012), https://doi.org/10.1088/1367-2630/14/1/013063.

4. Artur S. L. Malabarba et al., "Quantum Systems Equilibrate Rapidly for Most Observables," *Physical Review E* 90, no. 1 (July 2014), https://arxiv.org/abs/1402.1093.

5. Sheldon Goldstein, Takashi Hara, and Hal Tasaki, "Extremely Quick Thermalization in a Macroscopic Quantum System for a Typical Nonequilibrium Subspace," *New Journal of Physics* 17 (April 2015), https://doi.org/10.1088/1367-2630/17/4/045002.

6. Seth Lloyd, "Black Holes, Demons and the Loss of Coherence: How Complex Systems Get Information, and What They Do with It" (July 1, 2013), https://arxiv.org/pdf/1307.0378.pdf.

7. Paul Skrzypczyk, Anthony J. Short, and Sandu Popescu, "Extracting Work from Quantum Systems" (February 12, 2013), https://arxiv.org/abs/1302.2811.

Quantum Weirdness Now a Matter of Time

1. S. Jay Olson and Timothy C. Ralph, "Extraction of Timelike Entanglement from the Quantum Vacuum," *Physical Review A* 85, no. 1 (January 2012), https://doi.org/10.1103/PhysRevA.85.012306.

2. T. C. Ralph and N. Walk, "Quantum Key Distribution without Sending a Quantum Signal," *New Journal of Physics* 17 (June 2015), https://doi.org/10.1088/1367-2630/17/6/063008.

3. J. D. Franson, "Generation of Entanglement outside of the Light Cone," *Journal of Modern Optics* 55, no. 13 (2008): 2111–2140, https://doi.org/10.1080/09500340801983129.

4. Lucien Hardy, "Probability Theories with Dynamic Causal Structure: A New Framework for Quantum Gravity" (September 29, 2005), https://arxiv.org/abs/gr-qc/0509120.

5. Ognyan Oreshkov, Fabio Costa, and Časlav Brukner, "Quantum Correlations with No Causal Order," *Nature Communications* 3 (October 2, 2012), https://doi.org/10.1038/ncomms2076.

6. Lorenzo M. Procopio et al., "Experimental Superposition of Orders of Quantum Gates," *Nature Communications* 6 (August 7, 2015), https://doi.org/10.1038/ncomms8913.

7. Mateus Araújo, Fabio Costa, and Časlav Brukner, "Computational Advantage from Quantum-Controlled Ordering of Gates," *Physical Review Letters* 113, no. 25 (December 19, 2014), https://doi.org/10.1103/PhysRevLett.113.250402.

A Debate Over the Physics of Time

1. Julian Barbour, Tim Koslowski, and Flavio Mercati, "Identification of a Gravitational Arrow of Time," *Physical Review Letters* 113, no. 18 (October 2104), https://doi.org/10.1103/PhysRevLett.113.181101.

2. Sean M. Carroll and Jennifer Chen, "Spontaneous Inflation and the Origin of the Arrow of Time" (October 27, 2004), https://arxiv.org/abs/hep-th/0410270.

3. George F. R. Ellis, "The Evolving Block Universe and the Meshing Together of Times," *Annals of the New York Academy of Sciences* 1326 (October 13, 2014): 26–41, https://arxiv.org/pdf/1407.7243.pdf.

4. Maqbool Ahmed et al., "Everpresent Λ," *Physical Review D* 69, no. 10 (May 2004), https://arxiv.org/pdf/astro-ph/0209274.pdf.

A New Physics Theory of Life

1. Jeremy L. England, "Statistical Physics of Self-Replication," *Journal of Chemical Physics* 139, no. 12 (September 28, 2013), https://doi.org/10.1063/1.4818538.

2. Gavin E. Crooks, "Entropy Production Fluctuation Theorem and the Nonequilibrium Work Relation for Free Energy Differences," *Physical Review E* 60, no. 3 (September 1999), https://arxiv.org/pdf/cond-mat/9901352v4.pdf.

3. England, "Statistical Physics of Self-Replication."

4. Philip S. Marcus et al., "Three-Dimensional Vortices Generated by Self-Replication in Stably Stratified Rotating Shear Flows," *Physical Review Letters* 111, no. 8 (August 23, 2013), https://doi.org/10.1103/PhysRevLett.111.084501.

5. Zorana Zeravcic and Michael P. Brenner, "Self-Replicating Colloidal Clusters," *Proceedings of the National Academy of Sciences* 111, no. 5 (February 4, 2014), https://doi.org/10.1073/pnas.1313601111.

How Life (and Death) Spring From Disorder

1. R. Landauer, "Irreversibility and Heat Generation in the Computing Process," *IBM Journal of Research and Development* 5, no. 3 (July 1961), https://doi.org/10.1147/rd.53.0183.

2. Carlo Rovelli, "Meaning = Information + Evolution" (November 8, 2016), https://arxiv.org/abs/1611.02420.

3. Nikolay Perunov, Robert A. Marsland, and Jeremy L. England, "Statistical Physics of Adaptation," *Physical Review X* 6, no. 2 (April–June 2016), https://doi.org/10.1103/PhysRevX.6.021036.

4. Susanne Still et al., "Thermodynamics of Prediction," *Physical Review Letters* 109, no. 12 (September 21, 2012), https://doi.org/10.1103/PhysRevLett.109.120604.

5. E. T. Jaynes, "Information Theory and Statistical Mechanics," *Physical Review* 106, no. 4 (May 15, 1957): 620–630, https://doi.org/10.1103/PhysRev.106.620.

6. A. D. Wissner-Gross and C. E. Freer, "Causal Entropic Forces," *Physical Review Letters* 110, no. 16 (April 19, 2013), https://doi.org/10.1103/PhysRevLett.110.168702.

7. Andre C. Barato and Udo Seifert, "Thermodynamic Uncertainty Relation for Biomolecular Processes," *Physical Review Letters* 114, no. 15 (April 17, 2015), https://doi.org/10.1103/PhysRevLett.114.158101.

8. Jerry W. Shay and Woodring E. Wright, "Hayflick, His Limit, and Cellular Ageing," *Nature Reviews Molecular Cell Biology* 1 (October 1, 2000): 72–76, http://doi.org/10.1038/35036093.

9. Xiao Dong, Brandon Milholland, and Jan Vijg, "Evidence for a Limit to Human Lifespan," *Nature* 538 (October 13, 2016): 257–259, https://doi.org/10.1038/nature19793.

10. Harold Morowitz and D. Eric Smith, "Energy Flow and the Organization of Life" (August 2006), https://doi.org/10.1002/cplx.20191.

In Newly Created Life-Form, a Major Mystery

1. Clyde A. Hutchison III et al., "Design and Synthesis of a Minimal Bacterial Genome," *Science* 351, no. 6280 (March 25, 2016), https://doi.org/10.1126/science.aad6253.

2. C. M. Fraser et al., "The Minimal Gene Complement of *Mycoplasma genitalium*," *Science* 270, no. 5235 (October 20, 1995): 397–403, https://doi.org/10.1126/science .270.5235.397.

3. D. G. Gibson et al., "Complete Chemical Synthesis, Assembly, and Cloning of a Mycoplasma Genitalium Genome," *Science* 319, no. 5867 (February 29, 2008): 1215–1220, https://doi.org/10.1126/science.1151721.

4. D. G. Gibson et al., "Creation of a Bacterial Cell Controlled by a Chemically Synthesized Genome," *Science* 329, no. 5987 (July 02, 2010): 52–56, https://doi.org/10 .1126/science.1190719.

5. K. Lagesen, D. W. Ussery, and T. M. Wassenaar, "Genome Update: The 1000th Genome—A Cautionary Tale," *Microbiology* 156 (March 2010): 603–608, https://doi .org/10.1099/mic.0.038257-0.

Breakthrough DNA Editor Born of Bacteria

1. Jennifer A. Doudna and Emmanuelle Charpentier, "The New Frontier of Genome Engineering with CRISPR-Cas9," *Science* 346, no. 6213 (November 28, 2014), https:// doi.org/10.1126/science.1258096.

2. Motoko Araki and Tetsuya Ishii, "International Regulatory Landscape and Integration of Corrective Genome Editing into In Vitro Fertilization," *Reproductive Biology and Endocrinology* 12 (November 24, 2014), https://doi.org/10.1186/1477-7827-12-108.

3. Ruud Jansen et al., "Identification of Genes That Are Associated with DNA Repeats in Prokaryotes," *Molecular Microbiology* 43, no. 6 (March 2002): 1565–1575, https://doi.org/10.1046/j.1365-2958.2002.02839.x.

4. Eugene V. Koonin and Mart Krupovic, "Evolution of Adaptive Immunity from Transposable Elements Combined with Innate Immune Systems," *Nature Reviews Genetics* 16 (2015): 184–192, https://doi.org/10.1038/nrg3859.

New Letters Added to the Genetic Alphabet

1. M. M. Georgiadis et al., "Structural Basis for a Six Nucleotide Genetic Alphabet," *Journal of the American Chemical Society* 137, no. 21 (June 3, 2015): 6947–6955, https:// doi.org/10.1021/jacs.5b03482.

2. L. Zhang et al., "Evolution of Functional Six-Nucleotide DNA," *Journal of the American Chemical Society* 137, no. 21 (June 03, 2015): 6734–6737, https://doi.org/10 .1021/jacs.5b02251.

3. D. A. Malyshev et al., "A Semi-Synthetic Organism with an Expanded Genetic Alphabet," *Nature* 509 (May 15, 2014): 385–388, https://doi.org/10.1038/nature13314.

4. Georgiadis et al., "Structural Basis for a Six Nucleotide Genetic Alphabet."

5. Zhang et al., "Evolution of Functional Six-Nucleotide DNA."

The Surprising Origins of Life's Complexity

1. L. Fleming and D. W. McShea, "Drosophila Mutants Suggest a Strong Drive toward Complexity in Evolution," *Evolution & Development* 15, no. 1 (January 2013): 53–62, https://doi.org/10.1111/ede.12014.

2. Arlin Stoltzfus, "Constructive Neutral Evolution: Exploring Evolutionary Theory's Curious Disconnect," *Biology Direct* 7 (2012): 35, https://doi.org/10.1186/1745-6150-7-35.

3. M. W. Gray, "Evolutionary Origin of RNA Editing," *Biochemistry* 51, no. 26 (July 3, 2012): 5235–5242, https://doi.org/10.1021/bi300419r.

Ancient Survivors Could Redefine Sex

1. J. F. Flot et al., "Genomic Evidence for Ameiotic Evolution in the Bdelloid Rotifer *Adineta vaga*," *Nature* 500, no. 7463 (August 22, 2013): 453–457, https://doi.org/10.1038/nature12326.

2. D. B. Mark Welch, and M. Meselson, "Evidence for the Evolution of Bdelloid Rotifers without Sexual Reproduction or Genetic Exchange," *Science* 288, no. 5469 (May 19, 2000): 1211–1215, https://doi.org/10.1126/science.288.5469.1211.

3. D. B. Mark Welch, J. L. Mark Welch, and M. Meselson, "Evidence for Degenerate Tetraploidy in Bdelloid Rotifers," *Proceedings of the National Academy of Sciences* 105, no. 13 (April 1, 2008): 5145–5149, https://doi.org/10.1073/pnas.0800972105.

4. E. A. Gladyshev, M. Meselson, and I. R. Arkhipov, "Massive Horizontal Gene Transfer in Bdelloid Rotifers," *Science* 320, no. 5880 (May 30, 2008): 1210–1213, https://doi.org/10.1126/science.1242592.

5. Chiara Boschetti et al., "Biochemical Diversification through Foreign Gene Expression in Bdelloid Rotifers," *PLOS Genetics* (November 15, 2012), https://doi.org/10.1371/journal.pgen.1003035.

6. B. Hespeels et al., "Gateway to Genetic Exchange? DNA Double-Strand Breaks in the Bdelloid Rotifer *Adineta vaga* Submitted to Desiccation," *Journal of Evolutionary Biology* 27 (February 15, 2014): 1334–1345, https://doi.org/10.1111/jeb.12326.

Did Neurons Evolve Twice?

1. C. W. Dunn et al., "Broad Phylogenomic Sampling Improves Resolution of the Animal Tree of Life," *Nature* 452, no. 7188 (April 10, 2008): 745–749, https://doi.org /10.1038/nature06614.

2. Joseph F. Ryan et al., "The Genome of the Ctenophore *Mnemiopsis leidyi* and Its Implications for Cell Type Evolution," *Science* 342, no. 6164 (December 13, 2013), https://doi.org/10.1126/science.1242592.

3. Marek L. Borowiec et al., "Dissecting Phylogenetic Signal and Accounting for Bias in Whole-Genome Data Sets: A Case Study of the Metazoa," *BioRxiv* (January 20, 2015), https://doi.org/10.1101/013946.

How Humans Evolved Supersize Brains

1. Raymond A. Dart, "*Australopithecus africanus*: The Man-Ape of South Africa," *Nature* 115 (February 7, 1925): 195–199, https://doi.org/10.1038/115195a0.

New Evidence for the Necessity of Loneliness

1. G. A. Matthews et al., "Dorsal Raphe Dopamine Neurons Represent the Experience of Social Isolation," *Cell* 164, no. 4 (February 11, 2016), https://doi.org/10.1016 /j.cell.2015.12.040.

2. Pascalle L. P. Van Loo et al., "Do Male Mice Prefer or Avoid Each Other's Company? Influence of Hierarchy, Kinship, and Familiarity," *Journal of Applied Animal Welfare Science* 4, no. 2 (2011), https://doi.org/10.1207/S15327604JAWS0402_1.

3. Raymond J. M. Niesink and Jan M. Van Ree, "Short-Term Isolation Increases Social Interactions of Male Rats: A Parametric Analysis," *Physiology & Behavior* 29, no. 5 (November 1982), https://doi.org/10.1016/0031-9384(82)90331-6.

4. John T. Cacioppo et al., "Loneliness within a Nomological Net: An Evolutionary Perspective," *Journal of Research in Personality* 40, no. 6 (December 2006): 1054–1085, https://doi.org/10.1016/j.jrp.2005.11.007.

How Neanderthal DNA Helps Humanity

1. T. Higham et al., "The Timing and Spatiotemporal Patterning of Neanderthal Disappearance," *Nature* 512, no. 7514 (August 21, 2014): 306–309, https://doi.org /10.1038/nature13621.

2. S. Sankararaman et al., "The Date of Interbreeding between Neandertals and Modern Humans," *PLOS Genetics* 8, no. 10 (October 4, 2012), https://doi.org/10.1371/journal.pgen.1002947.

3. Matthias Meyer et al., "A High Coverage Genome Sequence from an Archaic Denisovan Individual," *Science* 338, no. 6104 (October 12, 2012): 222–226, https://doi.org/10.1126/science.1224344.

4. X. Yi et al., "Sequencing of 50 Human Exomes Reveals Adaptation to High Altitude," *Science* 329, no. 5987 (July 2, 2010): 75–78, https://doi.org/10.1126/science.1190371.

5. E. Huerta-Sánchez et al., "Altitude Adaptation in Tibetans Caused by Introgression of Denisovan-like DNA," *Nature* 512, no. 7513 (August 14, 2014): 194–197, https://doi.org/10.1038/nature13408.

6. David Reich et al., "Genetic History of an Archaic Hominin Group from Denisova Cave in Siberia," *Nature* 468 (December 23, 2010): 1053–1060, https://doi.org/10.1038/nature09710.

7. P. W. Hedrick, "Adaptive Introgression in Animals: Examples and Comparison to New Mutation and Standing Variation as Sources of Adaptive Variation," *Molecular Ecology* 22, no. 18 (September 2013): 4606–4618, https://doi.org/10.1111/mec.12415.

8. S. Sankararaman et al., "The Genomic Landscape of Neanderthal Ancestry in Present-Day Humans," *Nature* 507, no. 7492 (March 20, 2014): 354–357, https://doi.org/10.1038/nature12961.

9. B. Vernot and J. M. Akey, "Resurrecting Surviving Neandertal Lineages from Modern Human Genomes," *Science* 343, no. 6174 (February 28, 2014): 1017–1021, https://doi.org/10.1126/science.1245938.

10. M. Dannemann, A. M. Andrés, and J. Kelso, "Introgression of Neandertal- and Denisovan-like Haplotypes Contributes to Adaptive Variation in Human Toll-like Receptors," *American Journal of Human Genetics* 98, no. 1 (January 7, 2016): 22–33, https://doi.org/10.1016/j.ajhg.2015.11.015.

11. C. N. Simonti et al., "The Phenotypic Legacy of Admixture between Modern Humans and Neandertals," *Science* 351, no. 6274 (February 12, 2016): 737–741, https://doi.org/10.1126/science.aad2149.

The Neuroscience behind Bad Decisions

1. Ryan Webb, Paul Glimcher, and Kenway Louie, "Rationalizing Context-Dependent Preferences: Divisive Normalization and Neurobiological Constraints on Choice" (October 7, 2016), https://doi.org/10.2139/ssrn.2462895.

2. M. Carandini, D. J. Heeger, and J. A. Movshon, "Linearity and Normalization in Simple Cells of the Macaque Primary Visual Cortex," *Journal of Neuroscience* 17, no. 21 (November 1, 1997): 8621–8644, http://www.jneurosci.org/content/17/21/8621.

3. E. P. Simoncelli and D. J. Heeger, "A Model of Neuronal Responses in Visual Area MT," *Vision Research* 38, no. 5 (March 1998): 743–761, https://doi.org/10.1016/S0042-6989(97)00183-1.

4. O. Schwartz and E. P. Simoncelli, "Natural Signal Statistics and Sensory Gain Control," *Nature Neuroscience* 4, no. 8 (August 2001): 819–825, https://doi.org/10.1038/90526.

5. D. J. Heeger, "Normalization of Cell Responses in Cat Striate Cortex," *Visual Neuroscience* 9, no. 2 (August 1992): 181–197, https://doi.org/10.1017/S0952523800009640.

6. K. Louie, P. W. Glimcher, and R. Webb, "Adaptive Neural Coding: From Biological to Behavioral Decision-Making," *Current Opinion in Behavioral Sciences* 5 (October 1, 2015): 91–99, https://doi.org/10.1016/j.cobeha.2015.08.008.

7. K. Louie, M. W. Khaw, and P. W. Glimcher, "Normalization Is a General Neural Mechanism for Context-Dependent Decision Making," *Proceedings of the National Academy of Sciences* 110, no. 15 (April 9, 2013): 6139–6144, https://doi.org/10.1073/pnas.1217854110.

Infant Brains Reveal How the Mind Gets Built

1. Ben Deen et al., "Organization of High-Level Visual Cortex in Human Infants," *Nature Communications* 8 (January 10, 2017), https://doi.org/10.1038/ncomms13995.

2. Martha Ann Bell and Kimberly Cuevas, "Using EEG to Study Cognitive Development: Issues and Practices," *Journal of Cognition and Development* 13, no. 3 (July 10, 2012): 281–294, https://doi.org/10.1080/15248372.2012.691143.

3. Golijeh Golarai, Kalanit Grill-Spector, and Allan L. Reiss, "Autism and the Development of Face Processing," *Clinical Neuroscience Research* 6, no. 3 (October 2006): 145–160, https://doi.org/10.1016/j.cnr.2006.08.001.

Is AlphaGo Really Such a Big Deal?

1. David Silver et al., "Mastering the Game of Go with Deep Neural Networks and Tree Search," *Nature* 529 (January 28, 2016): 484–489, https://doi.org/10.1038/nature16961.

2. Volodymyr Mnih et al., "Human-Level Control through Deep Reinforcement Learning," *Nature* 518 (February 26, 2015): 529–533, https://doi.org/10.1038/nature14236.

3. Leon A. Gatys, Alexander S. Ecker, and Matthias Bethge, "A Neural Algorithm of Artistic Style" (September 2, 2015), https://arxiv.org/abs/1508.06576.

4. Christian Szegedy et al., "Intriguing Properties of Neural Networks" (February 19, 2014), https://arxiv.org/abs/1312.6199.

New Theory Cracks Open the Black Box of Deep Learning

1. Naftali Tishby, Fernando C. Pereira, and William Bialek, "The Information Bottleneck Method" (April 24, 2000), https://arxiv.org/pdf/physics/0004057.pdf.

2. Ravid Shwartz-Ziv and Naftali Tishby, "Opening the Black Box of Deep Neural Networks via Information" (April 29, 2017), https://arxiv.org/abs/1703.00810.

3. Alexander A. Alemi et al., "Deep Variational Information Bottleneck" (July 17, 2017), https://arxiv.org/pdf/1612.00410.pdf.

4. Pankaj Mehta and David J. Schwab, "An Exact Mapping between the Variational Renormalization Group and Deep Learning" (October 14, 2014), https://arxiv.org/abs/1410.3831.

5. Naftali Tishby and Noga Zaslavsky, "Deep Learning and the Information Bottleneck Principle" (March 9, 2015), https://arxiv.org/abs/1503.02406.

6. Brenden M. Lake, Ruslan Salakhutdinov, and Joshua B. Tenenbaum, "Human-Level Concept Learning through Probabilistic Program Induction," *Science* 350, no. 6266 (December 11, 2015): 1332–1338, https://doi.org/10.1126/science.aab3050.

A Brain Built from Atomic Switches Can Learn

1. Audrius V. Avizienis et al., "Neuromorphic Atomic Switch Networks," *PLOS One* (August 6, 2012), https://doi.org/10.1371/journal.pone.0042772.

2. Mogens Høgh Jensen, "Obituary: Per Bak (1947–2002)," *Nature* 420, no. 284 (November 21, 2002), https://doi.org/10.1038/420284a.

3. Per Bak, *How Nature Works: The Science of Self-Organized Criticality* (New York: Springer, 1996), https://doi.org/10.1007/978-1-4757-5426-1.

4. Kelsey Scharnhorst et al., "Non-Temporal Logic Performance of an Atomic Switch Network" (July 2017), https://doi.org/10.1109/NANOARCH.2017.8053728.

Clever Machines Learn How to Be Curious

1. Deepak Pathak et al., "Curiosity-Driven Exploration by Self-Supervised Prediction" (May 15, 2017), https://arxiv.org/abs/1705.05363.

Gravitational Waves Discovered at Long Last

1. B. P. Abbott et al., "Observation of Gravitational Waves from a Binary Black Hole Merger," *Physical Review Letters* 116, no. 6 (February 12, 2016), https://doi.org/10.1103/PhysRevLett.116.061102.

2. Rainer Weiss, "Electronically Coupled Broadband Gravitational Antenna," *Quarterly Progress Report, Research Laboratory of Electronics (MIT)*, no. 105 (1972): 54, http://www.hep.vanderbilt.edu/BTeV/test-DocDB/0009/000949/001/Weiss_1972.pdf.

Colliding Black Holes Tell New Story of Stars

1. H. Sana et al., "Binary Interaction Dominates the Evolution of Massive Stars," *Science* 337, no. 6093 (July 27, 2012): 444–446, https://arxiv.org/abs/1207.6397.

2. Krzysztof Belczynski et al., "The First Gravitational-Wave Source from the Isolated Evolution of Two Stars in the 40–100 Solar Mass Range," *Nature* 534 (June 23, 2016): 512–515, https://doi.org/10.1038/nature18322.

3. S. E. de Mink and I. Mandel, "The Chemically Homogeneous Evolutionary Channel for Binary Black Hole Mergers: Rates and Properties of Gravitational-Wave Events Detectable by Advanced LIGO," *Monthly Notices of the Royal Astronomical Society* 460, no. 4 (August 21, 2016): 3545–3553, https://arxiv.org/abs/1603.02291.

4. Pablo Marchant et al., "A New Route towards Merging Massive Black Holes," *Astronomy & Astrophysics* 588 (April 2016), https://arxiv.org/abs/1601.03718.

5. Doron Kushnir et al., "GW150914: Spin-Based Constraints on the Merger Time of the Progenitor System," *Monthly Notices of the Royal Astronomical Society* 462, no. 1 (October 11, 2016): 844–849, https://arxiv.org/abs/1605.03839.

6. Simeon Bird et al., "Did LIGO Detect Dark Matter?" *Physical Review Letters* 116, no. 20 (May 20, 2016), https://doi.org/10.1103/PhysRevLett.116.201301.

What No New Particles Means for Physics

1. Peter W. Graham, David E. Kaplan, and Surjeet Rajendran, "Cosmological Relaxation of the Electroweak Scale," *Physical Review Letters* 115, no. 22 (November 2015), https://arxiv.org/abs/1504.07551.

2. Nathaniel Craig, Simon Knapen, and Pietro Longhi, "Neutral Naturalness from Orbifold Higgs Models," *Physical Review Letters* 114, no. 6 (February 13, 2015), https://doi.org/10.1103/PhysRevLett.114.061803.

3. Nima Arkani-Hamed et al., "Solving the Hierarchy Problem at Reheating with a Large Number of Degrees of Freedom," *Physical Review Letters* 117, no. 25 (December 16, 2016), https://arxiv.org/abs/1607.06821.

The Strange Second Life of String Theory

1. ChunJun Cao, Sean M. Carroll, and Spyridon Michalakis, "Space from Hilbert Space: Recovering Geometry from Bulk Entanglement," *Physical Review D* 95, no. 2 (January 15, 2016), https://arxiv.org/abs/1606.08444.

A Fight for the Soul of Science

1. George Ellis and Joe Silk, "Scientific Method: Defend the Integrity of Physics," *Nature* 516, no. 7531 (December 18, 2014), https://doi.org/10.1038/516321a.

2. Joseph Polchinski, "String Theory to the Rescue" (December 16, 2015), https://arxiv.org/abs/1512.02477.

INDEX